Prospecting and Developing A Small Mine

Idaho Bureau of Mines Bulletin 20

Compiled by the staff of
Idaho Bureau of Mines and Geology

with an introduction by Kerby Jackson

This work contains material that was originally published by
the State of Idaho with the assistance of the United States Geological Survey.

This publication was created and published for the public benefit,
utilizing public funding and is within the Public Domain.

This edition is reprinted for educational purposes
and in accordance with all applicable Federal Laws.

Introduction Copyright 2014 by Kerby Jackson

Introduction

It has been over fifty years since the Idaho Bureau of Mines released its Bulletin 20, which was otherwise known as "Prospecting and Developing a Small Mine".

The author of this work, the late William Wesley Staley, was well known in those days for his expertise and his contributions to the mining industry, especially in Idaho. With his background as a mining engineer, he was especially suited to write a volume such as this one, and his concise and often straight forward writing style was especially appealing to information hungry small miners of his era. Not only were Staley's numerous publications on mining topics much sought after by miners who knew about his works, but also, many of his fans attributed much of their success at mining to him.

W.W. Staley was born on September 24th, 1898 in Singers Glen, Virginia. Despite this, he actually grew up in Portland, Oregon where he attended local schools. Though little is known of his earliest years, it is known that after graduating high school, Staley spent several years working as a chemist at a Portland cement plant.

In 1921, Staley enrolled at the New Mexico School of Mines and graduated in 1925 with a bachelor's degree in Mining Engineering. Not yet satisfied with his education, by 1931 he had also received a master's of science degree in metallurgy from the University of Idaho and a professional degree as an Engineer of Mines from the New Mexico School of Mines.

In 1930, William Staley joined the staff of the University of Idaho's School of Mines where he began to conduct research in mining and metallurgy. Up until his retirement in 1968, Staley authored numerous academic articles, textbooks and industry reference books on mining related subjects.

His academic work was so well received by the mining industry that he also acted as a consultant to some of the leading lead, silver and zinc mining companies in the United States, as well as worked as a consultant for the Atomic Energy Commission and the U.S. Bureau of Indian Affairs in the processing of mining claims held by several Indian tribes. Following his retirement from the University of Idaho in 1968, Staley moved to Tucson, Arizona where he split his time between consulting to the mining industry in that state and holding seminars on earth sciences at the University of Arizona until his death at the age of 93 in 1992.

In addition to this volume, some of William Staley's written contributions to the mining industry include such titles as "Mine Plant Design" (1936), "Introduction to Mine Surveying" (1939), "The Design of Small Wooden Headframes" (1937), "Timbering and Supports For Small Underground Mines" (1961), "Gold In Idaho" (1960), "Elementary Methods of Placer Mining" (1944) and twenty or more some others.

William Staley's written works, though now mostly retired to the dust bin of history, remain important because in large part, he wrote them with the small independent miner in mind and seemed acutely aware of his audience's often limited amount of engineering experience and capital. As a result, Staley's written works were of practical use to the small miner and the measure of their value seems only to increase as time goes on and this type of first hand knowledge becomes lost from generation to generation.

Despite the fact that we live in the so-called "Information Age" where information is supposedly only the push of a button on a keyboard away, true insight into age old mining techniques are illusive and often hard to come by, even to those of us who seek out this sort of information as if our lives depend upon it. Without this type of information readily available again, there is little hope that the average independent miner will ever have much of a shot at success.

This important volume and others like it, are being presented in their entirety again, in the hope that the average prospector will no longer stumble through the overgrown hills and the tailing strewn creeks without being well informed enough to have a chance to succeed at his ventures.

Kerby Jackson
Josephine County, Oregon
January 2014

ABSTRACT

Prospecting and developing a small mine requires much varied knowledge. The subject is introduced with a few remarks on the essential features of locating lode and placer claims, followed by discussion of sampling.

A working knowledge of geology, mineralogy, and rock types with their associations is important. Suggestions are offered for the economic location of surface openings. Because many inexperienced prospectors are unaware of the effect of topography (surface irregularities) on a vein's surface exposure, this subject has been explained and illustrated.

To guide the small operator, the essential features of the treatment of mine-run ore are briefly outlined. Methods, water requirements, and hand sorting are explained.

Very little prospecting can be accomplished without resorting to breaking and supporting ground. Hence, a few pertinent remarks are given on drilling and blasting, and on timbering small shafts and drifts.

During the development and mining period, costs become increasingly important. Recent data on the cost of equipment and supplies are offered to guide the small operator as the development program progresses. Because of their importance in this work, diamond drilling costs are given in some detail. Smelting costs and treatment schedules are included. Various payroll deductions are also included.

Services offered the prospectors by the Idaho Bureau of Mines and Geology and by several agencies of the Federal Government are explained.

A comprehensive bibliography on information generally of interest to the prospector and small operator concludes the bulletin.

INTRODUCTION

During the past several years an increasing number of requests about prospecting have come to the Idaho Bureau of Mines and Geology. Writers ask for information concerning the entire field of prospecting and small mine development. There is some indication of the revival of interest in prospecting and small operations as they were conducted before World War II. With this renewed interest in mind, the present publication has been prepared.

The procedure of prospecting with a view toward later development of a discovery into a profitable mine, is detailed. Because such a procedure requires knowledge of related and dissimilar topics ranging over a wide field, it is impossible to try and include all of the detail an inexperienced individual would need. He must seek additional information from reading the appropriate material listed in the Bibliography and supplement his reading by visiting operations of comparable activity to his own.

Unfortunately, many of the really helpful publications have been out of print for years and equivalent replacements have not been written. But the importance of accumulating a broad knowledge of mining geology through continual reading of related literature cannot be overstressed. And because mining meetings of local interest and magnitude usually produce beneficial ideas and information, attendance at these gatherings is also recommended.

Of ultimate importance to the small operator is the financing of the prospect. According to Wright (1935, p. 20) the following information should accompany proposals for financing:

1. Names and addresses of present as well as former owners of property.

2. Copies of title record and description of claims or parcels as given in office of recorder.

3. If property is held under partnership, copy of agreement; or if held under lease, copy of lease.

4. Complete statement of any mortgages or indebtedness against property.

5. Copies of available reports made by engineers or geologists.

6. Copies of available topographic and geologic maps of the district and a description of the geology of the district.

7. A recent engineer's report on the property which should include:

a. A description of the method used in sampling and name and address of assayer.

b. Detailed estimates of ore blocked out and assay results.

c. Estimate of probable and possible ore reserves and supporting data.

d. Detailed description of the surface and underground workings and their present condition and accessibility.

e. Inventory of buildings, plants, and equipment on the property.

f. Copies of laboratory tests or mill tests of the ore if any have been made.

g. Records of any ore shipments that may have been made, together with assays, results, and payments.

h. Information on source, kind, and amount of water supply and nature of water rights; on power supply and costs; on timber supply and costs; on labor supply and wage scale; on housing and living conditions; on transportation facilities and costs; on distances to nearest railroads; on conditions of roads; and all other information that may help in an estimation of production costs.

Because an understanding of proper sampling is basic to prospecting, that subject has been here reviewed in detail. Practical geological knowledge must be applied during the search for promising surface indications and their subsequent development. Therefore, this subject has like sampling, been presented extensively.

Although the suggestion about locating prospect workings has been repeated several times in the discussion on DEVELOPMENT, its importance warrants an introductory remark. The small operator should drive all of the early openings--shafts, drifts, tunnels, raises--in the vein. This procedure should be followed regardless of the resulting irregularities of the workings. And finally, too big a "bite" should not be attempted (length of surface tunnels should be limited to developing about 100 feet of vertical extent).

Mine-cost data depend on highly fluctuating factors--wages, equipment, supplies, taxes, demand, unit-output. For this reason, only a rather broad treatment was felt desirable. As a matter of fact, it may well be advisable not to let the restrictive influence of cost-keeping unduly dominate the prospecting or development program.

If greater detail is required about drilling, blasting, and ground support, the bibliography should be consulted.

To aid in obtaining out-of-print literature, a list of second-hand dealers is included. The larger city libraries will contain most of the books listed. Or a State University library may be consulted. The Engineering Societies Library, 29 West 39th Street, New York, New York, can provide, for a nominal sum, photostatic copies of almost anything. The Library of Congress, Washington, D. C. maintains a similar service.

LOCATION OF CLAIMS

TYPES OF CLAIMS AND PROCEDURE FOR LOCATING

The few suggestions offered will be confined mainly to locating on the public domain in Idaho (Federal land). There are two types of locations: lode and placer. For additional instructions the prospector should consult the mining laws of the state he is locating in and the Bureau of Land Management.*

LODE CLAIMS

The maximum size is 1,500 ft. long by 600 ft. wide with 300 ft. on each side of the vein.

When a discovery is made, a monument must be erected at the point of discovery: this procedure is prerequisite to locating a claim. The monument should contain a notice giving the locator's name, name of claim, date of discovery, and distance claimed along the vein each way from the monument. Within 10 days from date of discovery, the boundary of the claim must be marked by establishing a monument at each corner. At the time of marking the boundaries, a copy of the location notice must be posted at the discovery monument. This information must be included in the location notice:

1. Name of locator (or locators).

2. Name of claims (only one location to a location notice is permitted).

3. Date of discovery.

4. Direction and distance claimed along the ledge** from the discovery.

---

*In Idaho, a copy of the mining laws may be obtained from the State Mine Inspector, Boise, Idaho. The fee for the current issue is 25 cents. See also: Bureau of Land Management, Boise, Idaho, Circular No. 1941, on U. S. Regulations, and Ricketts, A. H., American Mining Law, Fourth Edition, Bull. 123, Calif. Div. of Mines, San Francisco. This excellent publication is issued periodically as a new edition with a new Bulletin number. However, the numbering of paragraphs remains unchanged. Unless specifically stated otherwise, all references to mining law herein will be to the Bull. 123 edition.

**Ledge, lode, vein are synonymous terms and have been held interchangeable.

5. The distance claimed on each side of the middle of the ledge.

6. The distance and direction from the discovery monument to such natural object or permanent monument that will fix and describe in the notice the location of the claim.

7. Name of mining district, county, and state.*

Monuments must be at least 4 in. square or 4 in. in diameter, and 4 ft. high. They should be trimmed so as to attract attention. A tree or post may be used. A pile of rock may be used.

Within 60 days following posting of the location notice, a shaft must be sunk on the lode. This shaft must be at least 10 ft. deep measured from the lowest edge and have an area of at least 16 sq. ft.

Open cuts and drill holes may be substituted for the shaft. The governing mining laws should be consulted regarding substitutions.

Within 90 days after location, a copy of the location notice must be filed for record in the office of the county recorder.

PLACER CLAIMS

The procedure for posting notices, size of monuments, and such detail is similar to a lode location. The size of the claim may not exceed 20 acres. There are restrictions governing the direction of the boundaries. On surveyed land, the boundaries must correspond to the land office survey.

Within 15 days after making the location, an excavation of not less than 100 cu. ft. must be made for prospecting the claim.

Within 30 days after location, a copy of the location must be filed in the office of the county recorder of the county.

ANNUAL ASSESSMENT WORK OR LABOR

Unless the current regulation is suspended by Congress, not less than $100 worth of labor must be done on each claim every year. At present (1961), the period ends each year at 12 o'clock meridian on September 1st. The form to be used is shown in the State Mining Laws.

A great many activities will satisfy these demands. The significance of the labor performed will depend on whether it is done within the claim or outside the claim.

*Printed forms suitable for this purpose may be obtained from printers in most mining areas.

Ordinarily, it is not difficult to prove that work within the claim benefits the claim. On the other hand, work outside the claim may bring about difficulties. It is best to consult a source like Ricketts, par. 484, 485, 486, in which are given many types of labor and expenditures that have been accepted as satisfactory by the courts

PATENTING THE CLAIM

The patent procedure is too involved and lengthy for discussion here. If patent is desired after the minimum improvements of $500 per claim have been made, the Bureau of Land Management should be consulted. There are both advantages and disadvantages to patenting the claim (Ricketts, 1943, par. 949).

LOCATING CLAIMS

It is beyond the purpose of this bulletin to discuss much further the ordinary details for locating mining claims. A few instances which commonly arise and about which many questions are asked will suffice.

Before examining this important and, in many respects, quite complicated and controversial subject, let it be reiterated that prospective locators of claims should obtain from the proper state agency a copy of the state's mining laws. Also, they should investigate <u>local</u> rules and regulations which may be in force in various mining districts. If there is no conflict with Federal or State statutes, the legality of local rules has been upheld by the courts.* The state publication usually contains a statement of the Federal statutes.

Three classifications of the domain (land) may be considered.

1. Public domain. Nearly all mining claim locations will be on National forest lands, although all of the states do not have land of this classification. For example, the Federal statutes do not apply in the original 13 states or parts thereof (Ricketts, 1943, par. 116). There are 17 so-called mining-law states (a somewhat modified statute applies in Alaska) in which the Federal mining laws are fully applicable (Ricketts, 1943, par. 114); each of the states has supplementary legislation.**

---

*In Idaho there are 183 organized districts. See Fed. Statutes, Title XXXII, Chapter VI, Revised Statutes, sec. 2324 or Dept. of Int., Bureau of Land Management Circular No. 1941.

**Idaho is one of the 17.

2. State land. The State of Idaho, and no doubt other states, provides legislation regulating the location of mining claims on state-owned land. In general, the procedure is similar to locating on the public domain. The prospective locator on state land in Idaho should obtain from the State Board of Land Commissioners, Boise, Idaho, a copy of the latest regulations. Some of the essential features are given in the Idaho Mining Laws.

3. Private Land. There is no provision by law for prospecting on private land. Unless mineral rights were specifically withheld when title was granted by the government (Federal or State--although the question usually arises about the right to prospect on what was originally a homestead grant) subsurface rights accompany the surface title. The existence of separate mineral and surface titles may be difficult to determine-- a comprehensive search of the title must be made. If prospecting on private land is contemplated, it is suggested that an agreement be reached with the owner before the investigation is begun.

Some undecided and some adverse opinions involve extralateral rights on locations adjoining homestead land. But these are too involved to receive further consideration here.

The previously stated size of a lode claim (600 ft. wide by 1,500 ft. long) represents <u>horizontal</u> measurements. An allowance for slope must therefore be made when locating on other than <u>flat</u> ground. For example, say that the surface slopes at about 25 degrees and the 1,500 ft. is measured on this sloping surface: the locator will have failed by 135 ft. to include the full length to which he was entitled. As the surface steepens, such an error would increase.

The statutes explain that the vein should pass through the middle of the end lines (600-ft. measurement) and the strike (course or direction) of the vein be parallel to the side lines (1,500-ft. dimension). Though location is perfectly valid if these regulations are not met, certain extralateral rights may otherwise be in doubt. Outcrops being as irregular as they are, one can not always conform to the requirements. Figure 1 illustrates diagrammatically the extralateral portion of the vein. During the early history of American mining, the prospector had insufficient time on many occasions to investigate properly his discovery before committing himself to a location. Consequently, many claims, improperly oriented, did not include the maximum amount of vein within the 600 ft. by 1,500 ft. Not only are very few outcrops exposed for more than a few feet without variations in strike, but equally important, the dip rarely shows in true direction or value. Because many early locations were made under pressure during the "rush," insufficient time was available to prospect properly the discovery and thus determine the true strike and dip. Today, the prospector is not so pressed by the mob panting at his back (although the uranium discoveries in the four-corner-states area apparently equalled the early gold rushes in this respect).

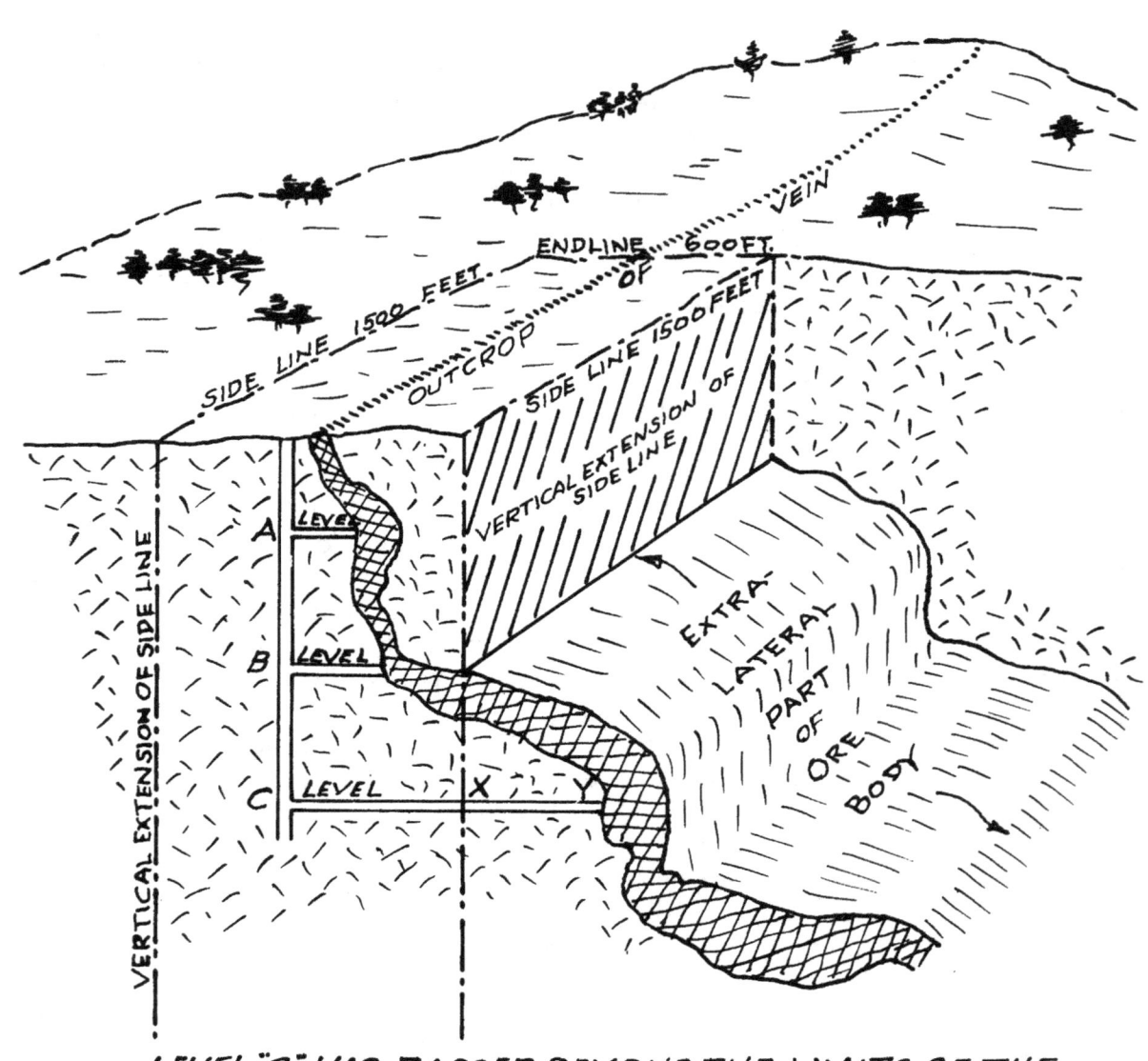

LEVEL "C" HAS PASSED BEYOND THE LIMITS OF THE CLAIM AT "X" AND IS TRESPASSING UNTIL IT REACHES "Y"

FIGURE 1

An inspection of Figure 2 will suggest how a claim might be wrongly oriented if the true strike is not first determined or at least approximated. The position of the outcrop is even more irregular when the vein has a flatter dip than used in the sketch or if the surface is more rugged.

The law (Federal law hardly more than suggests procedures, but the state laws usually go into detail) indicates the procedure for marking the boundaries (corners) of the claims. In concluding the discussion of this topic, one may remark that the monuments marking the boundaries of a claim prevail over other means of description (Ricketts 1943, par. 6, fn. 66).

The 600- by 1,500-ft. lode claim contains just over 20 acres.

Discovery or prospect tunnel

The Federal law provides that a tunnel location may be made for purposes of prospecting or discovery. The tunnel's length cannot exceed 3,000 ft. beyond its starting point. All undiscovered veins at the time of locating the tunnel and later intersected by the tunnel belong to the tunnel locator (Ricketts, 1943, p. 622, sec. 2323, and par. 725-728). The Federal law says little about the procedure to be followed for locating and identifying the tunnel location. This procedure has been left to the states. The California statutes are considered superior in this respect and are suggested as a minimum procedure (Ricketts, 1943, p. 655, par. 2308 and 2309; and par. 725, fn. 39). A brief statement of the California requirements are that a posted location notice at the point of commencement shall contain:

1. Name of locator.

2. Date of location.

3. Course of tunnel.

4. Description which ties tunnel to natural object or permanent monument.

5. Stakes on center line at intervals not exceeding every 600 ft. to the terminus 3,000 ft. from portal.

There are conflicting opinions about staking out a claim when the tunnel intersects a vein. This subject should be studied by prospective tunnel locator (Ricketts, 1943, par. 726 and p. 750 Form No. 54).

The tunnel location provides a means of prospecting an area either heavily timbered or covered with a thick soil mantle, or both. The Federal law provides that work must be "prosecuted with reasonable diligence for six months."

Mining tunnel or tunnel-right location

The purpose of this tunnel is only to gain access to and work a previously located mining claim. In the opinion of the mining claim owner, it is to his best advantage to enter the claim below the surface by means of a tunnel, because a right-of-way over another party's surface may be difficult to obtain. The entrance or portal of the tunnel must start either on land in possession of the operator of the mining claim or on public domain (Ricketts, 1943, par. CXCI and Idaho Min. Law, par. 47-1001, 1959 Ed.).

This brief discussion on mining law concludes with a reference to Figure 1. This illustration (Mining Truth, 1929) defines quite clearly the extralateral part of a vein. A term invariably used concurrently with extralateral is intralimital. This latter term defines the rights within the claim boundaries.

Consider the end view or vertical section. There we note that Level C departs from the claim at point X (which is on the side line). Extending C from X to Y is permitted with two restrictions:

1. The cross sectional size of X-Y must be confined to dimensions which only permit the passage of equipment. No crosscuts or lateral drifts may be driven.

2. The extension X-Y is open for inspection by the owner of the invaded ground.

One should also realize that once the extralateral extension of the vein is reached, all subsequent development (and this includes deadwork outside the vein) and mining must be confined to the vein. The only exception to this would be in the event the thickness of the vein became less than an economical mining width. (The narrowest minable width can vary for several reasons; probably at least 4 ft. would be acceptable). But considering the cross sectional view of the extralateral portion, we note that pursuing the vein within its confines results in an extremely irregular dip for winzes below C and inclined raises above C. Regardless of this uneconomical development planning, trespass will result if the country rock is penetrated after passing X (Ricketts, 1943, par. 553-555).

Permission must be granted at all times to the owner of the extralateral ground for inspection of work within his boundaries.

SAMPLING

Sampling the vein exposure is generally recognized as the best way to arrive at the value of exposed material. The procedure requires close attention to detail with careful and conscientious workmanship. The resulting values must be combined to obtain the weighted average.

Probably the main reason for sampling is to gather information to help classify ore exposures as to their several degrees of risk.

ORE CLASSIFICATION

There are many accepted definitions for classifying ore; each, in its own way, expresses the risk or degree of uncertainty involved. A simple and easily understood and applicable classification is appropriate here:

(1) Positive or proved ore: this is a block of ore developed on three or more sides (for example, length, height, width). The spacing between successive development openings permit little likelihood that the ore estimated is not present.

(2) Probable ore: ore developed on two sides is classified as probable ore (for example, length and width; or length and height).

(3) Possible or prospective ore: here the indication is exposed on only one side (for example, outcrop alone; lowest level in the mine).

Geological evidence, experience of the district, and depth of surrounding mines are among the criteria used to support all three classifications of ore; but the necessity of such support rapidly increases toward number (3). Actually, the entire future of a mine depends on the supporting evidence for possible ore. Only at mines with an extensive production history will (1) and (2) be present in great amounts at any one time. Evidence of positive and probable ore encourages financing and further exploration to convert possible ore into probable and positive ore. The prospector's efforts should be devoted toward quickly developing positive ore in substantial quantities (sufficient to equal any cash payment agreed upon). This necessity for rapid development means that all prospect workings should be driven in the vein. And these workings should be planned best to expose the vein on three, or at least, on two sides for tonnage estimates.

The three terms, positive, probable, and possible are similar; their meaning must not be confused. Under the section GEOLOGY APPLIED TO PROSPECTING AND DEVELOPMENT is described the geological knowledge needed to estimate possible ore.

TAKING THE SAMPLE

In Figure 2A the sample should be confined to the space between ADB. Drifts driven in the vein for the purpose of later sampling should avoid a rounded or arched top. If the top of the drift departs materially from a horizontal, square-corner type, the exposure will have to be represented by several samples. Or time will have to be spent in partially squaring up the drift. The sampled length of some portions will be disproportionate to the true length (W in Figure 2A). This disproportion will involve an error in reducing to the true length and proportionate weight of material to be removed. If the vein material is weak and almost immediate timbering is required, sampling should be done before placing the timber. To try to sample through lagging (or by removing the lagging) is very difficult and the results are uncertain. In the relatively shallow workings used for prospecting, arching the drift-back is of doubtful value, and an attempt to do so may actually result in substantial overbreak.

A sample along BEC could be considered. Theoretically, the result would probably equal ADB. However, the sample from this exposure would be longer and thus increase the uncertainty when reducing to the weight which is equivalent to the normal sample W. Also the lower part of the sample will not only be difficult to collect but possibly contaminated. A choice between the two areas to sample is much in favor of ADB. The sample taken there more closely approaches W in length, is less likely to be contaminated and is easier to remove and collect.

Before undertaking to remove the material, the area from which the sample will be cut must be thoroughly cleaned and a fresh surface exposed. This cleaning may be accomplished by scrubbing the surface with a wet wire brush; or in some instances, the condition of the area may require chipping away the old surface with the moil and hammer. The amount of attention devoted to this detail depends on the length of time the rock surface has been exposed to the action of moist mine air. It is not unusual to find a thick layer of fine material adhering to the walls of mine workings, and this coating may have become enriched by chemical action.

SALTING THE SAMPLE

Much has been written about salting or tampering with ore exposures. No doubt salting is or has been done on occasion, but a moment's reflection will suggest the difficulty of thus ordinarily influencing the results of a sample. It is highly improbable that salting an exposure containing other than gold, silver, or radioactive minerals could be accomplished to any degree. Seldom will the variation of a few tenths of a percent or even in excess of one percent of a base metal (iron, lead, zinc, copper, manganese, tungsten, etc.) greatly influence decisions about an early development program. On the other hand, it should be recalled that one ounce of gold or silver per ton is equal to about 0.0034+ percent. With the present price of gold a one-ounce ore has a value of $35; a variation of several tenths of an ounce would be quite important. A one-tenth-ounce change would equal 0.00034

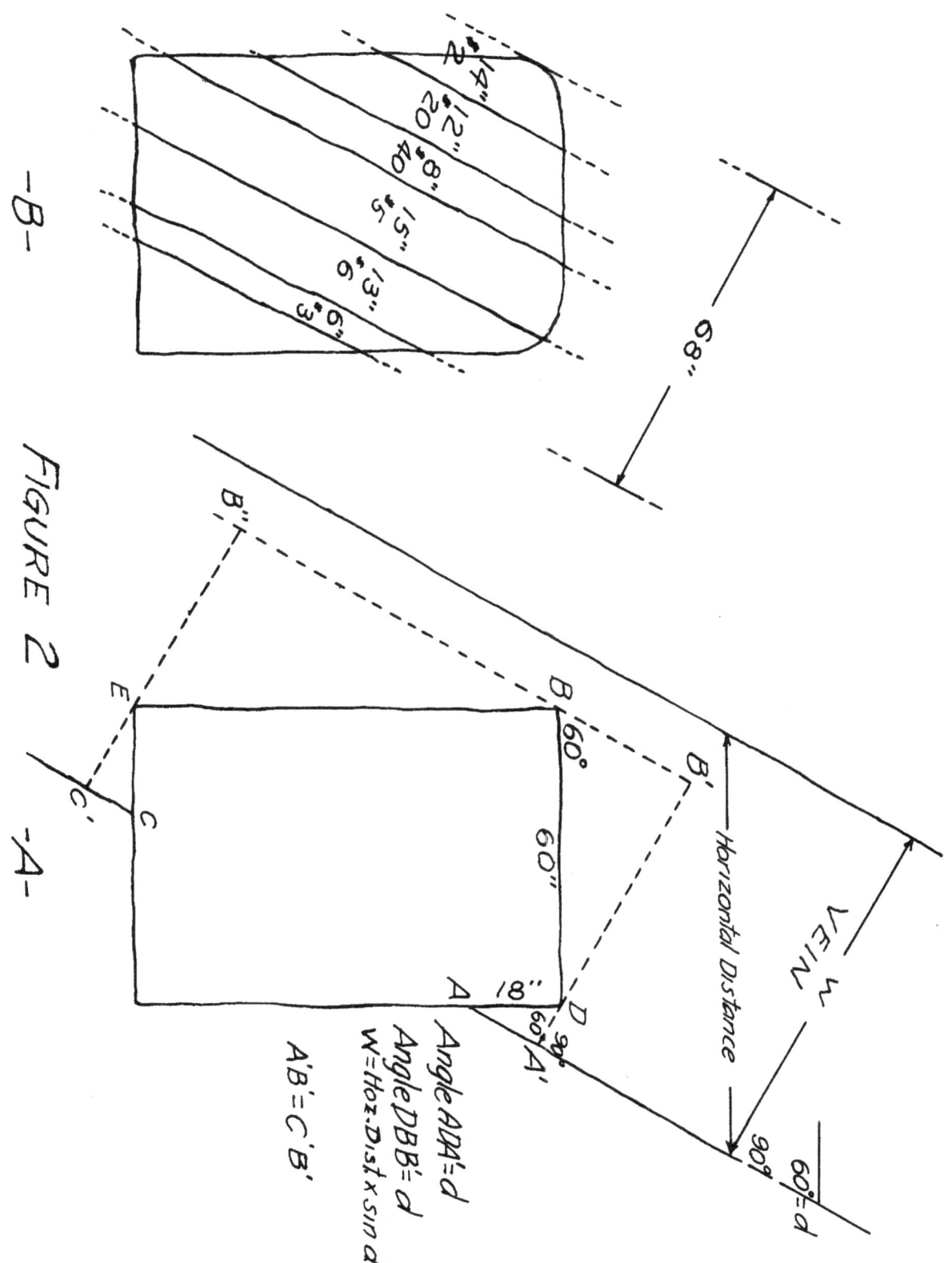

FIGURE 2

percent. Hence, tampering with a gold or silver ore would require the addition of only a very minor amount of metal. A sample may be salted intentionally or accidentally (carelessly or otherwise). Only because of the likelihood of accidental salting were the foregoing few remarks made. Careless sampling could influence even a base metal sample. More will be said about this later. If suspicious circumstances are encountered, the material removed when preparing the fresh face can be saved for a check assay. Only under most extraordinary conditions can tampering extend beyond the immediate surface of the exposure.

The true length to be assigned to the upper sample is A'-B'. If the wall is sampled, the effective length is B'-C'. The actual value of some of the dimensions may be difficult to determine; later averages and estimates of tonnage depend on these measurements; so the less complicated the measurements the more dependable the sample.

An important decision must be made before the actual sampling operation is undertaken. The dimensions of the sample to be cut or channeled along ADB must be such that the weight of material will correspond to the amount which would have been obtained along A'-B'. It is universally the custom to take A'-B' as the base. (Some companies use horizontal distances and reduce the final total to A'-B' with a factor).

AMOUNT OF SAMPLE

Although many experts have written on this subject, all of them, so far as I can determine, dismiss this important step with the remark that the weight must be proportional to the true length (length of influence) of the sample. (Weight here actually refers to the pounds or other units used; later, weight is used with a statistical meaning. It is quite important to distinguish between these two uses) It is suggested that the following procedure be used.

Let $ADB = x$ inches and $A'-B' = y$ inches

Also let P = pounds of sample that would normally be taken from $A'-B'$; then yP = in.-lb. product of material from the control sample. Therefore, the weight to be had from ADB would be $yP \div x$, pounds. (In practice, ADB is usually taken in several parts. See below).

Samples are usually cut with a hammer and moil. The hammer is a 4-lb. short-handed sledge-type. Commonly used is three-quarter-inch, hexagonal steel sharpened to a somewhat long point and tempered. The length of the moil is about 12 in.

COLLECTING THE SAMPLE

In collecting the cuttings, it is convenient to use two men. The helper holds an empty powder box (or similar sized receptable) close to where the sampling is in progress. An effort should be made to catch all of the cuttings. Any material which misses or bounces out of the box is <u>disregarded</u> unless a finely woven canvas cloth has been spread on the floor of the drift to catch this material; a rubber covered sampling sheet is best. All particles which fail to land on the sheet would be disregarded. The sheet must be thoroughly cleaned of fines and dust before starting a new sample. Especially is this true when sampling a gold or similar vein. If samples are to be taken by one man, the sampling sheet is necessary. In soft material, samples may be taken with a geology pick; the material is caught in a wide-mouthed sample sack held open with a heavy wire ring. Each sample sack should contain a tightly folded paper on which is recorded the number, date, and location of the sample. Other pertinent information may be recorded in the note book.

Samples are commonly removed in the form of either a channel or a V-notch. Some engineers go to excessive pains to keep the cut uniform in width and depth. Very likely there are instances when such extreme care is necessary. In general, however, a close and conscientious approach to uniformity should suffice. When cutting the sample, it is necessary that the hard, difficult portions not be slighted (they should contribute their fair share to the sample) and that the soft, easily obtained sections not be removed in excess. In other words, a reasonable attempt should be made to maintain the width and depth as first decided upon in spite of the occurrence of hard and soft spots. The dimensions of the cut should be such that the weight removed will reasonably represent the size or block (tonnage) to which it is referred. An exact recommendation for this does not seem to appear in the literature (Pierce and Kennedy, 1960, p. 28). (The authors suggest 6 lb. of coal per foot of thickness, this requires 12 sq. in. of channel per foot - 6" by 2"). I suggest that a sample weighing between 5 and 6 pounds per foot of sample cut, be taken as representative of the block. This amount would be the minimum: a sample of this weight will reduce to a channel-type of cut 4 in. wide, 1 in. deep, and 12 in. long. The units given refer to a sample taken normal (at right angles) to the dip--that is, A'-B'. The channel should be at right angles to the dip. If this angling cannot be done, and usually it cannot, (see Figure 2A, distance ACB) the measured length must be reduced to the equivalent true length, which may usually be obtained by measured distance (assuming this is horizontal) times the size of the dip angle, a formula that equals the true distance. Ordinarily, the back of a drift is horizontal, or sample cuts will be made horizontal if the normal measurements are inaccessible. When the measured distance exceeds the true distance, the weight (amount) of sample must be reduced in proportion to the ratio of the true distance to the measured distance. This reduction is most easily accomplished by changing the dimensions of the channel. Remember that the overall sample will usually be composed of several portions differing in their lengths and directions. This separation into several sections is deliberately done to simplify the sampling. Each such section must be reduced to true length and also the corresponding size of the channel determined. The sum of the various sections must equal the true length. It is not uncommon practice to "square-up" the corners of the sections to make their lengths and angles approach the normal or horizontal length and dip angle. Arching the back of a drift complicates these distances.

If we return to Figure 2A, the preceding discussion can be illustrated by assigning the following values in the figure: the calculations will be made on the basis of a cut 4 in. wide, 1 in. deep, and 12 in. long if cut normal to the dip. This cut is assumed to remove a 5-lb. sample of material weighing 180 lb. per cu. ft.

$$d = dip = 60 \text{ deg.}$$
$$AD = 18 \text{ in.}$$
$$BD = 60 \text{ in.}$$
$$A'DB' = BD \times \sin d + AD \times \cos d$$
$$A'DB' = 60 \text{ in.} \times \sin 60° + 18 \text{ in.} \times \cos 60°$$
$$= 60 \text{ in.} \times 0.866 + 18 \text{ in.} \times 0.5$$
$$= 52 \text{ in.} + 9 \text{ in.} = 61 \text{ in.}$$

Sample required = 5 lb. per foot of A'DB'
11.1 cu. ft. = 1 ton (assumed)
or 1 cu. ft. = 180 lb.

Portion AD

A'D = 9 in. and AD = 18 in.
9/18 x 5 lb./ft. = 2.5 lb./ft.
2.5 ÷ 180 = 0.0139 cu. ft.
0.0139 x 1728 cu. in./cu. ft. = 24.0 cu. in.
24 ÷ 12 in. = 2.0 sq. in. cross sectional area for channel cut.
If the sample is kept 1 in. deep,
   2.0 ÷ 1 = 2.0 in., the new width.
or if width is kept at 4 in., new depth = 2 ÷ 4 = 1/2 in.

Therefore, cross section of channel would be about 2 in. wide by 1 in. deep; or, 4 in. wide by 1/2 in. deep.

The choice between changing width or depth could be governed by distribution of minerals, variation in hardness, etc.

Portion BD

(52 ÷ 60) x 5 = 4.33 lb. per foot for section.
4.33 ÷ 180 = 0.024 cu. ft.
0.024 x 1728 = 42 cu. in.
42 ÷ 12 = 3.5 sq. in. for channel

Therefore, cross section of channel would be about 3.5 in. wide by 1 in. deep. These dimensions would give a total weight of material equal to a normal sample at 5 lb. per foot.

I am aware that the extensive literature on sampling contains many suggestions supported by mathematical proof (including the calculus) that attempt to demonstrate the necessity for various corrections to reduce measured distances to the normal or true distance. Such extreme accuracy would not only defeat the purpose of this paper but would be of doubtful value under any circumstances. The object here is to decide whether the prospect deserves further investigation or operation. This decision often approaches pure risk or speculation. A few percent variation in measurements should not unduly influence the decision. Indeed, at this stage, optimism is desirable.

SAMPLES FROM OTHER SOURCES

Introduction

Before beginning the concluding section on sampling--calculation of averages--a few words are required regarding samples from sources other than solid rock cuttings (sampling of bore holes and alluvial deposits require a little different approach). These additional methods may be grouped under--

(1) Grab samples:
    a. Consistently and fairly taken;
    b. "Hit or miss" sample, really a specimen.

(2) Metallurgical sampling:
    a. To control the milling or concentrating process;
    b. Smelter or reduction plant sampling.

A few comments about each will suffice; actually, those under (1) are of greater importance to the prospector.

Grab samples

Experience has shown that a handful or so of material taken during mucking out a round or drawing ore from a chute will, when similar samples are averaged over a period of time, check remarkably close to the precise mill sampling. From day to day, samples vary widely; but for a longer interval (say a month) the results are close. The amount taken for the sample should about fill a powder box for each 10 to 15 tons handled.

For use as a sample, the large piece or specimen is hardly worth the effort. The specimen has its purpose but not to represent tonnage. The resulting misrepresentation may just as easily under-value a prospect as over-value it. A specimen's acceptance as a sample must be avoided.

Metallurgical sampling

All properly designed concentrating plants have mechanically operated sampling devices. These sample cutters are designed to take a definite proportion of the material passing a particular point. Thus a very accurate sample is obtained which gives a reliable average for the mine-run ore (heads); the concentrates; the tailing; and for material at any other desired intermediate point. Ore entering the mill is either weighed by car or by a belt conveyor passing over an automatic weighing device. The separated products are also weighed. With this information--weights and samples--the metal content of a product at any stage is easily computed.

Ore arriving at a reduction plant is passed through the sampling house where an accurate weight and sample for assay are obtained. (The car or truck is weighed both loaded and empty). On the basis of this sampling, payments, penalties, and bonuses are allowed the shipper.

The preceding discussion should familiarize the prospector or small operator, if not re-assure him, with the sampling technique after his ore has been shipped; for it is not unusual for a difference of opinion to arise between the miner and the treatment plant over the metal content. Mine-run ore may be sent directly (possibly with hand sorting) to a custom mill or smelter. Sampling techniques at the small mine will seldom approach those practiced by the treatment plant. A natural result is for the miner to accuse the treatment plant of unfair practices. Careful investigation of such disputes suggest, however, that the trouble invariably lies in faulty sampling by the miner. If he takes a careful, representative sample; properly reduces it in size; and sends the sample to a reliable assayer; the result will closely check the smelter assay.

Let's see what is done with the 20 or more pounds taken as a sample.

The subsequent disposal of the cut or grab sample is no less important than the taking, averaging, and final assaying. All of the careful planning to obtain the sample goes for naught if proper treatment does not conclude the process. After obtaining the sample, the size of the larger pieces must be reduced. An exact figure for the extent to which the rough sample is reduced, is difficult to give. Actually, such a figure probably depends on the value of the material in the sample. A sample of gold ore or material of similar value would require more exact treatment than a sample of base ore. Recall that one ounce per ton of gold or silver ore is really only 0.0034+ percent of the ton of material. If a 20-lb. sample is used to represent one ton of a one-ounce ore, it would contain 0.01 ounces of the original gold. When this sample is finally reduced to the pound or so from which the one-ton (or less) portion is weighed for assay, it is really remarkable that any part of the original gold content finds its way to the final assay.

On the other hand, a lead, zinc, copper, or similar ore, will usually contain several percent of metal (an even larger bulk when one recalls that these are present as minerals and not as metals).

I have not intended to suggest that indifference or carelessness is of little consequence when reducing samples other than gold or silver (uranium should be included with these). On the contrary, the effort put forth should be commensurate with the value represented by the sample.

REDUCTION OF SAMPLE

If a small, laboratory-type jaw crusher, either mechanical or hand-operated is available, the reduction to maximum particle size is quickly and easily done. Thorough cleaning of the machine after crushing each sample must not be forgotten, especially in reducing any high grade material. If there is no crusher, the large pieces must be reduced with the sampling hammer. For an anvil a small piece of rail or steel plate may be used. This reduction must all be done on the sampling cloth and care taken that all broken material remains on the cloth. As in collecting the samples, any material that flies off the cloth is disregarded. The sample should be reduced to about one-half-inch maximum size (much smaller size is possible with the jaw crusher).

Following the preliminary reduction, the material is thoroughly mixed, generally by shoveling or scooping the outside edge of the material to a conical-shaped pile. For samples approximating 20 lb., opposite corners of the sampling cloth are lifted and the material rolled toward the opposite corner. This shoveling or rolling operation is continued until the sample appears uniformly mixed. With a reduced rolling action, the pile may be left in a nearly circular and conical heap. Next, the pile is flattened and divided into four quarters by marking with the moil. Opposite quarters are removed; the other two quarters are discarded. Rolling, quartering, and discarding are repeated until the original sample is reduced to a volume which may be held by the usual canvas sample sack (about 5 in. by 12 in. when empty). If a Jones splitter is at hand, the reduced material is passed through it with a saving of time and a more uniform sample removed. (The Jones splitter is a compartmented device which separates the sample into two equal portions by combining a number of narrow streams of material discharged on opposite sides of the splitter). To insure a maximum mineral content for the portion removed, many engineers modify the above technique by taking each half from the quartering procedure (or halves from the Jones splitter) and treating it as a separate portion. Each is mixed and quartered as before. Then halves from each of these are mixed. This quartering and mixing is continued until the desired bulk is reached.

To conclude the sample-reducing process, the operator grinds the contents of the sample sack to about minus 60- to 100-mesh in a fine pulverizer or with a bucking board and muller. In any event, the apparatus must be __thoroughly__ cleaned between samples. The finely reduced material is mixed by rolling on a rubber-surfaced sample cloth; it is then coned and a sample of several ounces sent to the assayer.

A final remark to conclude the subject of sampling: when one understands the reason for variations between samples, he also understands that the product from the most conscientious workman will more closely approach the exact content;

and the prospect deserves this same attention to detail.

For suggestions on sampling diamond drill core, drill cuttings (churn drill, rotary drill, etc.) and alluvial deposits, refer to the extensive literature on the subject. When the sample has been collected, many of the foregoing comments and suggestions can be applied (Hoover, 1909; Peele, 1941; Cummins, 1951 and Parks, 1957.

Part of the prepared sample should always be retained by the prospector from the shipment sent away for assay.

AVERAGING SAMPLES AND DETERMINATION OF TONNAGE

The operation of arriving at the average grade or value is known as finding the _weighted_ average of the samples. Note carefully that this procedure is not _weighing_ to represent quantity (pounds, etc.). Instead, it is statistical in effect, a process for obtaining an average by finding the effect of related units. A misunderstanding of these terms is widespread. Nothing is weighed. By combining the assay values (ounces, percent, cents, dollars, units, etc.) with the dimensions of the units or blocks of material (inches, feet, volume, weight, area) the true or weighted average is found. The other average commonly, and at times wrongly, used is arithmetical which is nothing more than the sum of the assay values divided by the number of assays. An arithmetical average can only represent the true average when each assay applies to an equal unit or weight of material.

A correct understanding of the weighted average is of prime importance, for on the basis of a correct average may depend the future of the prospect. And until a mining property produces material at a profit, it can be called by no other name. Therefore, we should examine, by means of simple arithmetic, the weighted average. At this point, it should also be remarked that the same procedure is followed to blend ore of different grades to produce a desired shipping or milling grade; to find the results of hand sorting; or for similar requirements. Units may be represented by any convenient combination of values and dimensions. The actual work is substantially reduced when the equivalent measurements defining several blocks can be made equal to each other.

Let,

W = length of sample (normal to the dip), in inches, feet, etc.

L = distance (interval) between samples, in inches, feet, etc.

D = height of depth or unit or block above or below sample, in inches, feet, etc.

V = assay value (ounces, percent, or currency value).

C = number of cubic feet per ton (if W L and D are in feet).

DIMENSIONS OF SAMPLE

How does one choose the dimensions? So far as W is concerned the previous discussion under sampling should, in part, suffice. It is not uncommon to increase W to equal the actual thickness of the vein, which requires judgment, for veins are far from having a uniform thickness.

The distance L between samples also requires thought and judgment. Too closely-spaced sample intervals result in an excessive sample and assay cost. On the other hand, samples taken too far apart may (and probably will) give a deceptive average. A spotty gold-quartz vein would require closer spacing than, say, a rather uniform, low-grade copper deposit or the common type of lead-zinc vein. It is worth taking the time first to make a preliminary geological (mineralogical) examination of the vein with a few test samples. These results will generally suggest a good sampling interval. The interval between samples is generally constant, but it should not be deliberately made so to simplify later office calculations. For example, some vein deposits (probably more often gold-quartz) have the disconcerting characteristic of more or less alternating high and low spots somewhat evenly spaced; this characteristic may occur both along either the strike or the dip, or both. Under such conditions, it would probably be fatal if the spacing consistently missed either the high or low values.

As used in calculating averages and volumes, L is expressed as the distance of influence. Accepted practice assumes that the value of a sample extends halfway to adjoining samples. Therefore, if $\underline{a}$ and $\underline{b}$ are the distances from one sample to each of two adjoining samples, L will equal (a/2 + b/2). If, as usually is the case, $\underline{a}$ equals $\underline{b}$, then the distance of influence corresponds to the sampling interval.

Values assigned to D, the third member necessary to calculate volume or tonnage, will generally require a knowledge of geology and the presence of development openings. These openings may be closely spaced tunnels below each other on the dip (certainly not in excess of 100 ft. apart); diamond drill holes; raises; or simply the assumption (based on geology and experience, either personal or of the district) that the vein will extend at least a certain distance above and below the sample. Without definite evidence to the contrary, this extension should seldom be assumed to exceed 25 ft. For example, a 7-ft. high drift in ore is to be sampled. With no contrary evidence from development openings, it may be assumed that the ore extended to 25 ft. above the drift and 25 ft. below the floor. To D would be assigned a value of 25 + 25 + 7 = 57 ft. Similar assumptions may be applied to W when the deposit is wider than the drift. More exact information is generally available for the width of the vein or deposit. But many replacement deposits in limestone or dolomite may be several hundred feet in width or length. To evaluate D properly, one should obtain the services of a competent mining geologist or engineer.

Values for V result from assaying the thoroughly mixed and reduced sample, which process has been explained. For calculating the weighted average, either the assay value (ounces per ton) or its monetary equivalent (dollars per ton) may be used. Because market quotations for nearly all metals and mineral substances change frequently, the assay value is probably preferable.

A determination of a reasonably accurate value for C is important; from this volume, tons of ore can be calculated. For a prospect or semi-prospect, the tonnage involved during its early life will not be large. Under these circumstances, a somewhat less exact value for C may be used without seriously compromising a final decision about future development. Actually, there would have to be quite a large change in the mineralogy of the ore to change C appreciably. A suitable method for assigning an approximate value to C will depend almost entirely on the skill and ability of the observer to estimate the mineral content of the ore--both proper identification and percentage of its minerals. Gold-quartz veins containing little or no sulfide minerals (pyrite, etc.) will not have to be estimated; the cubic feet per ton of quartz alone suffices. A 10 percent adjustment for voids and moisture should be applied. For such a procedure determine:

1. Gangue material: limestone, dolomite, quartz, granite, or whatever the rock may be. The gangue might even be a massive sulfide like pyrrhotite. Estimate the percentage making up the ore.

2. Mineral content: name of minerals and estimated percentage of each.

Then add 10 percent for voids and moisture.

(By recalculating a chemical analysis of the ore back to the corresponding minerals present--either by chemical calculations or using mineral compositions from a book on mineralogy--and by using each mineral's specific gravity, an accuracy exceeding practical considerations is easily obtained). Data necessary to find C are obtained by consulting various handbooks (Peele's Mining Engineers' Handbook in its various editions is recommended).

Here is an example to illustrate the steps to be followed. An inspection of a group of samples indicates the following percentage composition for an ore: limestone--60; quartz--15; galena--10; sphalerite--10; and pyrite--5. Assume ore has 10 percent voids and moisture.

From tables in a handbook or mineralogy book one discovers:

Mineral	Percent	Cu. Ft./Ton	Cu. Ft./Ton-Product
Limestone	60	11.9	714.0
Quartz	15	12.3	184.5
Galena	10	4.4	44.0
Sphalerite	10	8.0	80.0
Pyrite	5	6.3	31.5
	100	1054/100 = 10.54	1054.0

Add 10% for voids, water = 1.05

11.59 or 11.6 cu. ft. per ton.

Incidentally, the above procedure represents a good example of a weighted average.

If the weight per cubic foot is not shown in the table consulted, it may be found by multiplying the specific gravity of the mineral times 62.4 lb. (weight of 1 cu. ft. of water). This product divided into 2,000 lb. will give the cubic feet per ton. As an example, using quartz: 2.62 x 62.4 = 163.5 lb. per cu ft.; and 2,000 ÷ 163.5 = 12.2 cu. ft. per ton.

WEIGHTED AVERAGE

The component parts for obtaining the weighted average may now be expressed in formula form.

$$\text{Tons (or ounces) of metal} = \frac{W L D}{C} V$$

This formula is for one block. A number of blocks (decided by L, the spacing between blocks) are combined as follows:

$$\text{Total metal} = \frac{W_1 L_1 D_1}{C} V_1 + \frac{W_2 L_2 D_2}{C} V_2 + \frac{W_3 L_3 D_3}{C} V_3 + \ldots$$

Or,

$$\text{Total metal} = 1/C \ (W_1 L_1 D_1) V_1 + (W_2 L_2 D_2) V_2 + (W_3 L_3 D_3) V_3 + \ldots$$

In the absence of contrary information C is assumed constant and, in actual calculations, is applied to the equation only when the total is reduced to tons. For this reason, C will be dropped for the remainder of the explanation.

Letting V equal the weighted average, the formula is restated as follows:

$$V = \frac{((W_1 L_1 D_1) V_1 + (W_2 L_2 D_2) V_2 + (W_3 L_3 D_3) V_3 + \ldots)}{(W_1 L_1 D_1 + W_2 L_2 D_2 + W_3 L_3 D_3 + \ldots)}$$

That is, the sum of the assay-volume products (WLD times V) divided by the sum of the volumes (WLD) gives the weighted average assay. It should be apparent that only when the volume is combined with its respective value may the correct influence of each on the average be had.

In planning most practical sampling problems, several members may conveniently be made equal.

$$L_1 = L_2 = L_3 \ldots\ldots$$

$$D_1 = D_2 = D_3 \ldots\ldots$$

Such equalities are by no means exceptional. With a uniformly wide sampling face representing a vein of constant thickness,

$$W_1 = W_2 = W_3 \ldots\ldots$$

Usually, however, the W's are different. A common formula for V then becomes,

$$V = \frac{(W_1 V_1 + W_2 V_2 + W_3 V_3 + \ldots)}{(W_1 + W_2 + W_3 + \ldots)}$$

The product WV would be designated as assay-inches when W is measured in inches. When the W's are also equal, an arithmetical average results.

It must be added that making these dimensions equal is in no way "fudging." Sampling intervals are invariably made equal when the physical, mineralogical, and geological characteristics permit; the third dimension D, in any vein or ore body deserving the name, would not vary significantly over distances of a few feet.

Nothing has been really said as to actual value for L. In practice it varies from 5 to 20 feet when sampling along a drift, stope, raise, and sundry openings. But on the surface, drill holes may be separated by 100 ft. or more on certain deposits. At least one case is on record where a distance of over 1,000 ft. between samples proved satisfactory. A correlation with excellent geological data made this spacing possible. As the arrangement of values in the ore shoot approaches a homogeneous mixture, the spacing may be increased.

The data in Table 1 will illustrate the calculations for a weighted average. The sampling data necessary for calculating V are given in columns 1, 2, 3, 5, and 7.

DISCUSSION OF TABLE 1

In column 4, the distance between each sample is shown because it is universally accepted that the effect of a sample extends halfway to the adjoining samples; for sample 10 the 6% is effective halfway back toward sample 9 and halfway forward to sample 11. The total effective distance, the sum of these two partial distances, is as shown in column 4, equal to 8/2 + 12/2, or (8 + 12)/2, which equals 10; this result is recorded as L. A similar procedure is followed to get the distance of influence for the remaining samples. In column 6, the three dimensions for each block are combined to give the volume. The assay-volume product is given in column 8 (column 6 x column 7). In the bottom line of the table the averages and totals are given. Column 2 is averaged by dividing the sum of the sample lengths by the number of samples. The result equals the average width, which is used later for calculating the tonnage. The totals of columns 3 and 4 should check each other. These sums equal the length of the block for which the average is found. (In the present example, the overall length of the block could have been extended beyond sample 15 a distance determined by one's familiarity with the persistence of the ore shoot).

Inasmuch as column 8 represents the product of (W L D) times V, it follows that the average V will result from dividing the sum of column 8 by the sum of column 6. The result is the weighted average shown on the bottom line of column 7. The value is 9.39 percent.

The tonnage for the above example may also be computed. With the exception of C all of the required data are given. A value for C would be determined by the method previously suggested. For this example, assume C to be 11 cu. ft. per ton.

Column 3 gives the length of the block as 70 ft. and column 2 gives the average width as 4.83 ft.

Then,

$$4.83 \times 70 \times 50 = 16,905 \text{ cu. ft.}$$

$$\text{weight} = \frac{16,905}{11} = 1,537 \text{ tons}$$

$$\text{metal} = 1,537 \times 9.39\% = 144.3 \text{ tons}$$

ADDITIONAL EXAMPLES OF WEIGHTED AVERAGE

Two remaining examples involving weighted averages should be brought to the prospector's attention. Not unusually, the width (thickness) of an ore shoot or vein will be less than can be mined without breaking an appreciable

Table 1--Calculating the Weighted Average

(1)	(2)	(3)	(4)	(5)	(6)	(7)	(8)
Sample No.	Length of cut, W, ft.	Distance between samples, ft.	Distance of influence L, ft.	Penetration, D, ft.	WLD volume cu. ft.	Assay V, %	Assay-vol. product, (WLD) V
9	--	--	--	--	--	--	--
10	4	8	(8+12)/2 = 10	50	2,000	6	12,000
11	5	12	(12+12)/2 = 12	50	3,000	9	27,000
12	6	12	(12+20)/2 = 16	50	4,800	12	57,600
13	5	20	(20+12)/2 = 16	50	4,000	11	44,000
14	5	12	(12+10)/2 = 11	50	2,750	7	19,250
15	4	10	(10+0)/2 = 5	50	1,000	5	5,000
Avg. or Total	$\frac{29}{6} = 4.83$	8/2+12+ 20+12+10 = 70	70	$\frac{300}{6} = 50$	17,550	$\frac{164,850}{17,550} = 9.39$	164,850

amount of waste wall rock. At one time wages and output were conducive to hand-sorting to remove the waste and leave it behind as fill in the stope. Mining costs today rarely permit this method. It is cheaper to concentrate a lower grade mine product. Too many small operators fail to realize the effect on average grade of breaking a narrow vein to a "mining width." They have sampled the vein and obtained a gross value per ton; but during the mining a few inches to several feet of wall rock is unavoidably broken to provide working space. Or, the miner gets careless and shoots down waste; and sometimes a weak wall fails in spite of utmost care. In any event, the whole product is shipped on the assumption that each ton corresponds to the vein assay. In one of the southern counties of Idaho, two small-scale operations were closed down because of a related incident. Local men had a profitable market for limestone to a nearby sugar beet refinery. But they got careless and included too much aplitic dike rock--a fine grained granite dike that upon casual inspection closely resembled the limestone--in one too many shipments. For several years now, that refinery has bought its limestone in Nevada. Other interests in the same general locality worked up a market with a roof-covering manufacturer for fine mica. The first car load of granite with some mica was the last!

What is the resulting grade of 2.5 ft. of vein assaying 15 percent lead when the vein is broken to include 1.5 ft. of barren wall rock?

This question suggests that about 4 ft. is the minimum mining width. Miners have tried to drill in a stope 12 to 18 in. wide using a 2 in. by 12 in. plank as a platform. Such a narrow stope may be economical under some extreme conditions, but generally 4 ft. is taken as the minimum width. Very high grade ore could influence the minimum width either way, depending on management's policy. A low-grade vein would reduce the mining width to the bare minimum.

But to return to the question:

  2.5 ft. times 15 percent   = 37.5 ft.- %

  1.5 ft. times 0   = 0 "

and $\frac{37.5}{4}$ = 9.37

The calculations show that the material shipped would contain only 9.37 percent lead.

Examples of this kind are almost endless. There is no way to estimate the tons of rock containing a few inches of high-grade ore, all of which has mistakenly been shipped for ore. Yet the smelter is invariably accused of sharp practices.

One final example to conclude the discussion on sampling: it frequently happens that the vein or lode exposure in the face of a drift or across a stope is composed of many well-defined and distinguishable parallel stringers. Some gold-quartz veins of this type have a wide variation in the gold content from stringer to stringer. Several ages of deposition may be represented. It is not unusual for certain periods represented in the vein to be totally lacking in values: or for the values to differ greatly between earlier or later injections that finally formed the present vein.

For reasons clear only to prospectors, certain stringers will be sampled and assayed. Very encouraging results are obtained. The rest of the stringers are ignored but the entire face is assumed to assay equal to the sample. Even a casual inspection of the various quartz veinlets suggests a difference in their texture which generally also means a change in gold content. And just as important, the inspection will show the presence of wall rock between the quartz veinlets.

Figure 2B is a sketch illustrating the face of a drift in which is exposed a series of gold-quartz veinlets separated by less valuable material. The problem is to find the average grade from hanging wall to footwall.

```
  14 in.  x  $ 2  =  28 in. -$
  12 in.  x    20  = 240    "
   8 in.  x    40  = 320    "
  15 in.  x     5  =  75    "
  13 in.  x     6  =  78    "
   6 in.  x     3  =  18    "
  68 in.       $759 in. -$
```

Average = 759/68 = $11.16 per ton.

Assuming that the two outside stringers can be left in place:

```
  12 in.  x  $20  = 240 in. -$
   8 in.  x   40  = 320     "
  15 in.  x    5  =  75     "
  13 in.  x    6  =  78     "
  48           $713 in. -$
```

Average - 713/48 = $14.85 per ton.

Before the final decision is made to discard the outside 20 inches (their combined value is $2.30 per ton) mining, shipping, and treatment costs must be determined. Probably it will prove cheaper to include the low-grade stringers. The careful drilling and blasting required to leave them in place would probably reduce the tons-per-man-hour output.

ALLUVIAL SAMPLING

Many other applications of the weighted average to sampling illustrate the same underlying principle in all cases. One, however, merits attention: alluvial or placer deposits. In this type of operation, the units are generally resolved to cents per cubic yard. Linear dimensions result from the depth or partial depth of the opening and some method of using the distances between openings. Alluvial or flat-lying deposits are investigated by using drill holes, pits, or trenches. The expression is generally "cents-feet product" or "cents-yards product." The volume of the hole or pit is customarily constant throughout the depth and samples of the material are taken at uniform intervals (every foot or so). If these factors are not constant, the weighted average of the opening must first be found before the opening is combined with adjacent holes.

Many patterns (triangles, quadrilaterals, polygons, all of which may be regular or irregular) may be assumed for combining the holes with each as a focal point for surrounding holes. When trenches are cut across the deposit, the area of each vertical section is determined (usually a trapezoidal formula is applied) and the cents-square-yard product extended halfway to the adjoining trenches. With the drill holes, patterns of triangles, regular quadrilateral, irregular polygons, etc. may be tried. The lines connecting the holes may extend directly from hole to hole, or more complicated figures may be arranged. In general, the more sides involving the multiple use of the holes the figure has, the more refined will be the average. The literature dealing specifically with alluvial sampling should be consulted (Jackson and Hedges, 1939; Parks, 1957). Electronic computers have been programmed for use in reducing these complicated figures to the average (Krumlauf, 1960).

GEOLOGY APPLIED TO PROSPECTING AND DEVELOPMENT

IMPORTANCE OF GEOLOGICAL KNOWLEDGE

If the prospector or small producer expects to realize the maximum benefits from his efforts, he should have a practical working knowledge of geology as it is applied to mining. Several excellent and very readable books have been written for this express purpose. But unfortunately, most have been out of print for many years. In the Bibliography, additional comments will be made on this subject.

The available geological literature--from federal, state, and private publishing companies--is extensive, but little is written with the prospector specifically in mind. Noteworthy exceptions are listed under FURTHER RECOMMENDED READING (Farrell and Moses, 1912; Gunther, 1912; and Spurr, 1926.) But even the language and terms of such publications are technical and much of the discussion is controversial in style, which is confusing to those not familiar with the fundamentals. Without a broad preparation of reading and training, the casual or practical student is soon discouraged or hopelessly bewildered by the maze of conflicting theories.

Even the most elementary discourse on mineral deposits must pre-suppose some familiarity with minerals, rocks, and geological features, which simply means that to a limited extent the common minerals, rocks, and geological features can be recognized and understood. A brief review of an extremely broad and complex topic follows:

MINERALS

This subject is so extensive that the thought of attempting a brief explanation is appalling. Anyone planning on prospecting should obtain an elementary text on mineralogy (Loomis, 1948; Golden Press, 1957) and a suite of about 100 minerals and rocks.* Peele (either of the three editions) contains an excellent section on mineralogy, rocks, and geology of interest to prospectors. The FURTHER RECOMMENDED READING list suggests others.

Minerals occur in every conceivable degree of color; they range in hardness from very soft (talc) to extremely hard (diamond); their densities may range from less than that of water (1.0) to that of native gold (about 19.0). Almost every one of the chemical elements contributes to some mineral formula. Of several thousand distinct minerals, however, less than 100 have commercial importance. However, as science develops, elements occurring as traces in other commercially important minerals come into demand. In this respect the metal cadmium, which is recovered from zinc smelting,

---

*Wards Natural Science, 302 N. Goodman Street, Rochester, New York.
 Denver Fire Clay Co., Denver, Colorado
 Many State Bureaus of Mines have available for a small fee elementary sets of common minerals typical of their state.

may be mentioned. Its presence in zinc ores is minor; another such element is hafnium which occurs with zirconium in the mineral zircon.

With a reasonable degree of accuracy, the presence of certain minerals may be used to predict the extent of mineralization. Such minerals, listed in Table 2 (Farrell and Moses, 1912, p. 132-133 and Bateman, 1950, p. 20-21) are known as primary (hypogene-high temperature) minerals. It is sufficient here to confine the extent of Table 2 to the four ranges of temperature and pressure shown. The column headed Contact Metamorphic also includes many of the minerals deposited under the higher temperatures and pressures of cooling magmas and pegmatites. It will be noted that, while there is broad overlapping, it is not impossible to estimate closely the presence or absence of the four zones of a vein. This interpretation should by no means be used to suggest that all four of the zones were originally formed. None or all of these may have been formed; erosion easily could have removed one or all of them; or conditions for confining temperature and pressure may not have been ideal for the zoning effect. There is ample evidence that the zoning effect is extended horizontally as well as vertically. The low temperature zone is named the epithermal; intermediate zone, mesothermal, and high temperature zone, hypothermal. The minerals listed in Table 2, both ore and gangue, are known as primary, meaning that they are original constituents of ore deposits and not secondary (formed from the original or primary minerals by alteration, for example, by descending ground water).

ROCKS

Classification

For the purpose of this bulletin, rocks may be broadly classified as igneous, sedimentary, and metamorphic. The igneous rocks are classified in several ways: according to origin (intrusive or extrusive), acid or basic nature, mineral content, and texture. It will satisfy our purpose to confine the discussion to naming the more common rocks generally associated with, or causing, or influencing the formation of mineral deposits.

Table 2

Primary Minerals Which May Classify Ore Deposits by Zonal Arrangement

Ore Minerals		Contact Metamorphic	High Temp.	Intermediate Temp.	Low Temp.
Argentite[1,5]	(silver[6])				a
Arsenopyrite	(arsenic[6])	b	b	a	b
Bornite	(copper[6])	b	b	a	b
Chalcocite[1]	(copper[6])		b	b	b
Chalcopyrite	(copper[6])	b	b	a	b
Cinnabar	(mercury[6])		b	b	a
Cobaltite	(cobalt[6])			b	a
Enargite	(copper[6], arsenic[6])			a	b
Galena	(lead[6])	(a)	(a)	a	(a)
Gold[1]		b	b	b	a
Jamesonite				b	a
Magnetite	(iron[6])	a	a	(b)	
Marcasite[1,2]	(iron[6])			b	a
Molybdenite	(molybdenum[6])	b	b	b	b
Orpiment	(arsenic[6])				a
Polybasite[5]					a
Pyrargyrite[5]					a
Pyrite[1,3]	(iron[6], sulfur[6])	b	a	a	b
Pyrrhotite[4]	(iron[6], sulfur[6])	b	a	b	
Realgar	(arsenic[6])				a
Ruby silvers (pyrargyrite, proustite)					a
Silver[1]				b	a
Specularite	(iron[6])	a	a	c	
Sphalerite (zincblende) (zinc[6])		a	a	a	b
Stephanite[5]					a
Stibnite	(antimony[6])				a
Tetrahedrite	(copper[6], silver[6])		b	a	b
Zincblende (sphalerite) (zinc[6])		a	a	a	b
Gangue Minerals					
Adularia				(b)	a
Albite		a	a	a	
Alunite					a
Amphibole		a	a		
Barite			b	b	b
Chalcedony				(b)	a
Calcite		b	b	b	b

Table 2 (Cont'd.)

Carbonates (calcite[1], dolomite[1])	b	b	b	b
Chlorite (high iron)		a	b	
Chlorite (low iron)			b	a
Diopside	a	b		
Dolomite		b	b	b
Epidote	a	a		
Fluorite	a	b	b	b
Garnet	a	a	c	
Hornblende	b	b		
Marcasite[1,2]			b	a
Muscovite	b	b	b	
Pyrite[1,3]	b	a	a	b
Pyroxene	b	a		
Pyrrhotite[4]	b	a	b	
Quartz[1]	b	b	b	b
Rhodochrosite				a
Rhodonite	a	a	a	a
Sericite	b	b	a	a
Siderite[1]		b	b	a
Tourmaline	b	a		
Tremolite	a	b		
Vesuvianite	a			
Wollastonite	a			

[1] Also may be secondary.
[2] Also gangue
[3] May be ore: gold bearing; copper bearing.
[4] May be nickel bearing; platinum bearing; copper bearing.
[5] Not common ore mineral, but presence is good indication of zone.
[6] Important metal in mineral

a - Characteristic of zone.
b - May be present   (in fact, it is quite likely).
c - Presence unlikely.
( )- Uncertain

Igneous rocks

1. <u>Extrusive</u>. Extruded from central vents, or fissures--lava flows: basalt, rhyolite, trachyte, latite, andesite; fragmental deposits: tuff, volcanic breccia.

2. _Intrusive_. Intruded into earth's crust; may be revealed by later erosion or earth movement: granite, monzonite, diorite, dacite, gabbro, diabase, peridotite. They are acid (granitic) to very basic (peridotite). The acid or granitic rocks are granite to diorite with a great variety of subdivisions and names depending on mineral and free-quartz content. Facts suggest that many ore deposits are related to granitic rocks (Emmons, 1937; Hulin, 1945). Intrusive rocks are generally coarser grained than extrusive rocks.

    a. _Dikes_. Tabular intrusive bodies that cut across the stratification or bedding of the intruded rocks are called dikes. Dikes differ widely in thickness (from almost microscopic to hundreds of feet thick), length, and vertical extent. In many instances they appear to be closely associated with mineral deposits. Dikes may themselves become the host rock for mineralization.

    b. _Sills_. These sheet-like intrusive bodies occur between the bedding or stratification. Under some conditions sills have acted like dams or impervious obstacles and have thus confined mineralization to the rocks below the sill.

    c. _Stocks, batholiths, etc._ These large masses, most of them granitic, are exposed through erosion. The intrusion of these masses may have caused extreme metamorphism and deformation of the intruded and overlying rocks. Ample evidence exists of the close association between ore deposits and the activity accompanying and resulting from these massive intrusions. Two famous ones are the Idaho batholith occupying a large part of central Idaho, and the Boulder batholith of western Montana, which contains the famous Butte area.

    d. _Pegmatites_. These are very coarsely crystalline dikes, commonly of granitic composition. The feldspars, quartz, and mica crystals may reach many feet in size. Pegmatites are the source of many minerals (mica, feldspars, heavy metal oxides, uranium minerals).

Sedimentary rocks

Sedimentary rocks are derived from the erosion or decomposition, or both, of all rock-types with later redeposition. They are the limestones, dolomites, sandstones, shales, conglomerates, sands, gravels, clays, arkose, etc. There are many gradations between them. Under the right conditions many, if not all of them, have acted as host rocks for mineral deposition. The limestones and dolomites may contain siliceous areas or layers and chert nodules, and may range in color from white to black. In some instances--for example, the Bayhorse, Idaho, area-- the ore shows a preference for the dolomitic limestone (this is a limestone containing several percent of dolomite); this preference is apparently not uncommon. With several exceptions, the most important lead-zinc deposits in the world are in limestone-dolomite-type rocks. (a famous exception is the Coeur d'Alene district of Idaho where the deposits are in Precambrian quartzitic rocks with comparatively minor carbonate rocks).

Metamorphic rocks

These rocks are formed by the action of heat, pressure, and chemical solutions on igneous and sedimentary rocks. There is a great variety of metamorphic rocks and the exact origin of many of them is a subject for conjecture. It will satisfy our purpose to consider only a few of them.

1. Gneiss is a coarse-grained rock in which bands rich in granular minerals (like quartz and feldspar) alternate with bands in which schistose minerals (like mica) predominate.

2. Schist is a medium or coarse-grained rock with subparallel orientation of the micaceous minerals that dominate its composition. Usually it is named from some predominating mineral--mica schist, garnet schist, hornblende schist, etc.

3. Quartzite is metamorphosed sandstone.

4. Slate, derived from shale or clay, is a fine-grained rock having well-developed fissility or cleavage.

5. Marble is metamorphosed limestone or dolomite.

6. Phyllite is a mica-bearing rock intermediate between slate and schist.

RELATION OF MINERAL DEPOSITS TO CERTAIN GEOLOGICAL FEATURES

General statement

In most mineral deposits one of the essential requirements is that there have been adequate openings in the rock to permit the ore minerals to be deposited; or an opening from which replacement can start. Therefore, the following discussion will indicate the importance of igneous activity and of structural (earth) deformation in creating the openings necessary in the formation of ore deposits. Several other guides useful in the search for ore deposits will be discussed.

Igneous intrusive bodies as guides

Many mining areas have been centers of intense igneous intrusion and extensive earth disturbances. Granitic intrusive bodies ranging from small stocks to large batholiths cause extensive fracturing of the invaded (overlying) rocks, which may be igneous, sedimentary, or metamorphic. Periodic resurging of the intrusion, followed by cooling, results in fracturing which may not infrequently extend into the intrusion itself. It is not uncommon for the overlying rocks to have been faulted and folded before intrusion. Tension and compression openings, fissures, shear zones, and similar openings in almost every conceivable pattern, may be formed during intrusion.

In many ore deposits dikes are important. It is important to determine if the dikes are pre-ore or post-ore. If the dikes are pre-ore then the relation of the dike to the ore should be studied. In this way certain guides to the ore may become apparent. If the dikes are post-ore then the knowledge that there is a lack of a relation between the ore and the intrusive dike is of equal importance.

An example of the relation of igneous intrusion to ore deposits can be seen in the Idaho batholith. A map containing the outline and boundaries of the batholith (including numerous "islands" of invaded sedimentary, metamorphic, and older igneous rocks), shows a linear contact of about 2,400 miles (Staley, 1960, p. 12). If one goes a step farther and locates on this map known prospects, mines, and other determinable indications of prospecting (there are hundreds of them!), he will see that Emmons' (1937) suggestion that the bulk of mineralization will be within a mile or so of the granite-invaded rock contact, is supported to an astonishing degree. Having compiled such a map of Idaho (unpublished), I find this proposition to be true of Idaho deposits. And interestingly enough, I noted localities of strong surface mineralization with no evidence of intrusive igneous rocks. This pattern suggests the presence of such rocks at depth. In certain localities, Emmons' theory does not hold. But where exploration has been done, intrusive rocks have been encountered at relatively shallow depths. The peripheral contact of the Idaho Batholith with all of its irregularities offers an extensive area for prospecting.

Fractures as guides

The most common earth disturbance related to ore deposits is fracturing (formation of faults and joints). No definite and convincing conclusions may be drawn to relate ore deposits to any particular fracture-type, although numerous examples exist of mineralization occurring in or in conjunction with normal, reverse, and thrust faults. Fracturing forms the channelways for entry of the ore-bearing solutions; the receptables for ore-deposition; and the starting-places for replacement. Indeed, practically all deposits are directly or indirectly related or associated with fracturing. The shape and form of fractures varies greatly from sub-microscopic in width and length to fractures that are many yards in width to many miles in length.

The direction of movement along a fault should be carefully worked out to determine where the larger openings will be expected to occur in relation to the dip and strike of the structure. In general on a normal fault the larger openings will be along the parts of the fault that have the steeper dips whereas on a reverse fault the larger openings will be on the flatter parts of the fault.

The concentration of ore at the intersection of two or more fractures is a very common occurrence. In some districts where the ore is localized by an intersection the immediate junction is barren. In this case, the concentration generally occurs at a distance from the intersection. In general, the greater concentration of ore occurs where the fractures intersect at an acute angle. There is no positive assurance that intersections will have a concentration of ore; however, they are well worth testing.

Folds as guides

Many ore deposits occur in folded rocks and in some particular part of a fold. By fold is meant anticlines, synclines, rolls, drag folds, cross folds, bands, domes, and monoclines. Mineralization may take place at the top, upper flank, lower flank, anticlinal or synclinal part, or crest or trough of the fold. Examples of locations in every possible part of a fold may be quoted. Moreover, there are numerous examples of the solutions choosing small folds within larger folds.

Many folds are younger than the ore. If this is the case the ore will have taken on the contorted condition of the enclosing rocks. In many cases, the ore will thin on the flanks of the folds because of the squeezing effect.

Where the folds are older than the ore, the ore may bear no relation to the folds. However, commonly the ore will follow the folding. More important, in this case, is that in the process of deformation, fracturing resulted which produced openings. Fracturing which resulted from the folding generally occupies a characteristic position on the fold. As a result the fractures can be predicted with considerable accuracy. Accompanying the folding was intense squeezing which may have produced impermeable layers which hampered the flow of the ore solutions or may have acted as traps in which the ore was deposited.

Physiographic guides

Topography is one of the most useful guides to ore deposits. In some areas the veins are visible for miles because they are more resistant to weathering and stand as a ridge. Broken Hill in Australia is a classic example. In other areas, the mineralized zone is more easily weathered and as a result appears as a depression. Unfortunately where the mineralization occurs in a depression it is generally covered with alluvium (stream deposits) or talus (gravity fall debris). Under these conditions it is necessary to study the float on the downslope side of the depression.

The manganese deposits near Deming, New Mexico were discovered by a study of the float on the lower edge of a depression.

Physiographic features useful as guides are by no means always present. In the Coeur d'Alene district many of the best veins do not show on the surface.

Mineralogical guides

The minerals that are in an ore deposit are probably the most significant guide. Certain minerals have an affinity for other minerals. The association of molybdenum with the chalcopyrite in certain copper deposits is a good example. The complexity of mineral associations is beyond the scope of this paper and the reader is referred to the FURTHER RECOMMENDED READING list at the end of this paper.

Certain alteration products are associated with certain types of deposits. The following table (Schwartz, 1939, p. 181) is given as an example:

With hypothermal mineralization: garnet, amphiboles, pyroxenes, tourmaline, biotite.

With mesothermal mineralization (and also in many deposits classed as hypothermal and epithermal): sericite, chlorite, carbonates, and silica.

With epithermal mineralization: some sericite, often much chlorite and carbonate, adularia or alunite

Pyrite is the most important indicator that mineralization has occurred. The presence of pyrite indicates that sulfur has been introduced into the rocks and because most of the important ore minerals are sulfides the pyrite confirms mineralization. The presence of pyrite, however, does not imply that an ore deposit has been formed but is merely a good guide. In addition to the introduction of sulfur, pyrite also indicates the introduction of iron. Alteration of the iron is the cause of the formation of limonite and hematite which give the gossan its red-brown-yellow color. This discoloration of the rocks is a good prospecting guide.

No report of this kind is complete without mention of the importance of quartz as a guide to mineralization. However, quartz--as in the case of pyrite--is of very little use by itself; other criteria must accompany it. All prospectors know of the association of gold with quartz.

Rock-type as a guide

Rock-type preference for mineral deposition is commonly spoken of as favorable and unfavorable. Generally rocks that fracture to many small angular pieces (breccia) and thus present maximum area of contact for the solutions are most favorable. But the rocks in such small fragments must not readily alter or decompose to an inactive and impervious clay or gouge. The hard, brittle rocks with minimum alteration are ordinarily most favorable. But there are many exceptions. Lead- and zinc-carrying solutions show a marked preference for replacing carbonate rocks (calcite, siderite, limestone, dolomite, dolomitic limestone); and copper solutions commonly react with the siliceous rocks. Examples are known where this choice is made even when limestone and quartz are in opposing contact with each other in the fissure.

In Table 3 is given a generalized review of the rock choice made by mineralizing solutions. These examples are but a few typical cases (condensed from tables of Rock Types in Newhouse, 1942).

Certain impervious layers (sills, shale beds, silicified tuff beds) or dikes may act as dams and hinder or prevent the passage of solutions. Deposition or replacement reaches its maximum in the rocks on the entering side.*

Table 3

Metal and Associated Rock Types

Metal	Favorable Rock	Unfavorable Rock	Remarks
Gold	Rhyolite and andesite	Trachyte	Brittleness and shattering important
Gold	Hard, brittle, porphyry sills, brittle limestones and quartzites	Soft shales	
Gold	Granodiorite	Schistose lava flows	Granodiorite more brittle

---

*Hanover and Kelly, New Mexico; Leadville and Camp Bird Mine, Colorado, may be mentioned. I suspect this same kind of damming action has also occurred in the Bullion Gulch area east of Hailey, Idaho. Here a basal conglomerate in the Wood River Formation may well have confined the mineralized zone to the Milligen Formation, letting only a very minor part pass into the overlying Wood River. Comparative production in Blaine County offers strong support for this idea. The Milligen in Blaine County has grossed at least $30,000,000 in lead-zinc-silver; whereas the Wood River shows only about $5,000,000, part of which is from gold-quartz veins in the intrusive rocks. Also at Hanover, New Mexico, the bulk of the ore is found on the west side of the dikes. But just enough was found on the other side to require that both sides be prospected.

Table 3 (Cont'd.)

Gold	Conglomerate	Fine grained tuff	Conglomerate fractures better
Gold	Massive andesite or dacite; conglomerate and graywacke	Tuffs, slates, lavas, carbonates, schist	
Gold	Graywacke, greenstone	Slate	
Gold	Arkose	Graywacke	Graywacke became schistose
Gold	Quartzite	Shales	More open fissures in quartzite
Gold	Quartzite	Limestone, schist	Small fractures disappeared in limestone and schist
Gold	Granodiorite	Dikes of rhyolite porphyry	A change in physcal properties of rock shifted emphasis
Gold	Dikes of rhyolite porphyry	Granodiorite	
Silver (gold)	Diabase or other brittle rock		
Silver, lead, gold	Sedimentary rocks, sandstone, argillite, shale		Larger openings in sedimentary rocks
Silver	Limestone, and quartzite; conglomerate	Sandstone and shales	
Lead-zinc	Silicated bands in limestone		
Lead-zinc	Dolomites and limestones	Shaly layers	Better fracturing in carbonate rocks
Zinc	Crystalline limestone	Blue limestone	

Table 3 (Cont'd)

Molybdenum	Albite granite	Schist, metamorphosed sediments	Unfavorable rocks could not maintain openings
Mercury	Sandstone	Shale	Sandstone fractured
Copper-silver	Tuff and breccia	Shale	Veins pinch out in shale
Copper (gold and silver)	Volcanics	Slaty tuffs	
Copper	Andesite volcanic breccia	Massive andesite	
Copper	Thick bedded limestone	Shaly limestone	
Tin	Schist	Quartzite and limestone	Pegmatites were in schist

Here are several idealized examples (many actual cases could be cited) that may be of some practical significance. A not uncommon occurrence is lead-zinc sulfides (galena and sphalerite) with pyrite replacing limestone (copper-iron sulfide minerals may also be present--chalcopyrite, bornite). Assuming erosion has not been too extensive, evidence of the following rearranging of the lead and zinc may be expected (there are numerous actual examples of such occurrences.*) Oxidation of the sulfides by oxygen-carrying, descending, surface water will leave a porous, brown-stained outcrop or gossan indicative of the action. The brown color results from the iron oxides (and complex iron-lead-zinc silicates). The lead sulfide is altered to insoluble, or nearly insoluble, lead sulfate with little or no change in its position. This alteration is the mineral anglesite, which, because of its relatively high insolubility, remains in place. It is not uncommon to find a small core of galena in the center of the anglesite pseudomorphs. Soluble zinc sulfate, formed from the sphalerite, is carried downward by the descending solutions. When the solutions come in contact with the limestone, zinc carbonate is formed as smithsonite. The physical appearance of this replacement in many instances so nearly resembles the original

---

*Hanover and Kelly, New Mexico; Mackay, Idaho; Nicholia, southeast Lemhi County, Idaho; Colorado; Utah; etc.

limestone as to defy casual recognition. At Kelly, New Mexico this replacement was overlooked for many years and the zinc carbonate was thrown on the waste dump.

Below the water table (and this depth is arbitrary because of fluctuations of ground water level) the oxidation ceases and primary sulfides are found in their original condition and position. If sufficient time elapses and conditions are right, the anglesite is converted to the stable lead carbonate (cerrusite). Under the proper environment (presence of silica), the zinc may be altered to a number of complex zinc silicate minerals (calamine, etc.)--called by the prospector oxidized zinc-- and may remain with or just below the lead. Gold in this type of deposit is rare. Practically all of the silver will remain with the lead. Many deposits contain in the oxidized zone a section of complex lead-zinc-iron silicate material.

Summarizing the foregoing, one finds:

1. At the surface, a leached, brownish colored, porous outcrop (gossan);

2. lead sulfate and lead carbonate in and below the gossan;

3. possibly zinc silicates (or iron-lead-zinc silicates) below or with the lead mineral;

4. below the silicates or lead minerals, the zinc carbonate grading into primary, unaltered sulfides;

5. depending on the intensity of the activity, a more or less barren gap between 3 and 4;

6. below the secondary zinc and the water table, the primary sulfide zone and in certain cases an enriched zone where the descending solutions deposited their metallic lead.

Only under very sparsely mineralized conditions is evidence of the primary zone missing, that is, was the oxidation complete.* The above described process is not confined to a vertical deposit. It has occurred with relatively small dips.

A second instructive occurrence is the separation of gold and silver with deep redeposition and enrichment of the gold. The presence of manganese oxides or heavy stain in the outcrop of a vein would suggest further investigation for enrichment areas. Through chemical processes, the pyrite in the vein is oxidized and its sulfur converted to sulfuric acid and other necessary chemicals. Either the rocks or the surface waters contain traces of chlorides. These chlorides, from their reaction with the acid and the manganese oxides, form soluble gold

---

*At Nicholia, Idaho, sulfides seem to be absent although a very large body of of silicate and carbonate zinc remains. Also at White Knob, near Mackay, Idaho, sulfides seem to be absent from some of the occurrences of oxidized zinc.

chloride. Through the medium of the descending waters, the gold chloride is carried to a lower horizon. Insoluble silver chloride remains behind. The gold-bearing solutions above or near the water table (200 to 800 feet have been reported: Emmons, 1917, p. 314) are decomposed and native gold enrichment results. Because this process of dissolving, transporting, and precipitation of metallic gold is repeated many times, there may be several enriched zones.

To summarize the discussion on gold deposition, one may say that the presence of manganese in a gold-silver-pyrite-quartz vein suggests the likelihood of silver and gold enrichment zones.

A final, typical occurrence will explain the formation of the secondary (supergene) enrichment of certain copper deposits. Examples of these deposits occur in New Mexico, Arizona, Utah, and Nevada, as well as many other places in the world. Such deposits are the major sources of copper and produce, in addition, a significant amount of molybdenite. The gold production of the United States fluctuates widely as these mines vary in their production.

The rock in which the deposits occur is commonly granitic in character: granite, granodiorite, monzonite. There are some exceptions: for example, in the Globe-Miami, Arizona area, the host rock is the Pinal schist; and in the Santa Rita, New Mexico, deposit the change from igneous rock to limestone is so gradual that differentiation between the two is extremely difficult. These masses (several thousand feet wide and equally as long may even approach several thousand feet in depth) originally contained less than 0.01 percent primary copper. Pyrite was more or less homogeneously scattered throughout the rock. Action of oxygen-carrying, descending surface water oxidized the pyrite to iron oxide and formed the typical brownish gossan cover. The acid and other chemicals resulting from this action, dissolved the copper. As the transporting solutions moved downward, the copper was deposited on pyrite grains and other receptive nonsulfide minerals. The redissolving, transportation, and deposition was repeated many times. As this process developed, local areas of high grade azurite and malachite (copper carbonates), chryscolla (copper silicate), the several copper oxides, and the masses of native copper were formed. Rare rich sulfide pockets will also occur. The result, as these deposits are found today, is several hundred feet of typically brown, leached zone or gossan; and several hundred feet to over a thousand feet or more of enriched copper sulfide (chalcocite). This latter zone grades into the original material (protore), and the grade ranges from more than one percent at the top of the enriched zone to the original amount. The depth of the minable enriched zone depends on what percentage management can profitably operate. At present, for open pits, this figure seems to be about 0.4 percent. For an underground operation, it is about 0.6 to 0.7 percent.

A brownish to almost black, porous outcrop is probably the best single indication of mineralization; this indication is not confined to any particular metal. Nearly all of the big and famous deposits--high grade, low grade, lode, and massive-type-- were overlaid by a gossan. Gossans should always be prospected.

RADIOACTIVE MINERALS*

As is true of many other mineral deposits, radioactive minerals are found in both primary and secondary occurrences. They are found both in place and in placers.

Massive primary mineralization is mostly associated with granitic rocks and with pegmatites. However, primary minerals are found in the sandstone deposits of the Colorado Plateau.

Through the alteration of the primary minerals, secondary minerals also occur in the same rocks. But the most extensive occurrence of secondary minerals is in sedimentary rocks: sandstones, shales, limestones, phosphate rocks, asphaltic material, etc. The states of Colorado, Utah, New Mexico, and Arizona are well known for such deposits.

There is a widespread occurrence in alluvial (placer) deposits. Monazite especially, is a common placer mineral. Nearly all Idaho placers carry monazite to a greater or lesser extent. It has been suggested that an even distribution of the monazite-bearing gravels in South Korean placers would provide a mantle one foot thick over that entire country! Some such statement might well apply to Idaho (Staley, 1948; Savage, 1961). In placers, only the more resistant minerals persist: euxenite, fergusonite, samarskite, and columbium and tantalum.

Primary radioactive minerals occur under temperature and pressure conditions similar to those for other intermediate zone minerals (lead, zinc, copper, silver) (Nininger, 1954, p. 27). A few uranium minerals found under high temperature conditions associated with ilmenite and some zirconium minerals, are usually found in pegmatites.

Prospecting for radioactive material is done with either the Geiger counter or the scintillation counter.**

---

* Largely derived from Nininger, 1954.

** Uranium Prospectors' Handbook. Repro-Tech. Inc., 3535 Tejon Street, Denver 11, Colorado (1954)

Prospecting for Uranium, Supt. of Documents, Washington 25, D. C. (1949). Price 30 cents.

blank

PROSPECTING

OUTLINE OF PROCEDURE

Of the many procedures that may be followed in prospecting, this bulletin concerns itself with development after the preliminary search has indicated an outcrop or deposit of likely value.

An outline of customary prospecting methods for such a purpose may be reviewed as follows:

- A. Actual search should be made by walking over the ground and looking for outcrops (indicated by projections above or shallow depressions below the surface; or brownish-colored leached areas) and float; by inspecting mounds of earth left by ants and burrowing animals; by observing the preponderance of any particular species of vegetation (sage brush, mesquite, pinon, cactus, etc.); and by noting stained or other discolored areas (iron or manganese).

- B. Search may be made in areas surrounding operating mines for appropriate signs listed here.

- C. Areas having rock characteristics similar to those of known districts should be searched. Examples of such relations are rock associations with certain metals (their ores); for example, lead-zinc, usually in limestone or dolomitic limestone; gold in quartz veins or gold carrying pyrite in granitic rocks or quartz veins; copper in limestone or so-called porphyry (usually a granite-type rock); coal or petroleum in sandstones, shales, limestones.

- D. Areas of extensive structural disturbances should be investigated: folds, faults, shearing, igneous activity. For a careful study, apply geological techniques given in a previous section.

- E. Geophysical prospecting may be used.

    1. Surface: though a great many methods beyond the scope of this paper may be used, two are within the capabilities of the prospector.

        a. Magnetic methods for buried placer channels or magnetic deposits: magnetite, pyrrhotite, ilmenite, possibly chromite may be discovered by use of dip needle, superdip or magnetometer.

b. Radioactive methods for uranium minerals and thorium minerals: either or both the Geiger counter and the scintillation counter may be used.

   2. Airborne: both magnetometer and scintillometer are widely and successfully used as airborne methods.

F. Trenching should be used. With the availability today of the bulldozer, trenching and road building are fast, and compared with the old pick and shovel method, reasonably low in cost. This machine is commonly used to expose the outcrop. Under many circumstances, it has limited the necessity for a tunnel claim. The amount of useless trenching may be materially reduced by first determining the general strike and outcrop position of the local vein system (see Fig. 3). Trenches should be as near at right angles to the strike as possible.

G. Pits or test shafts may be useful. In the past these openings have been generally used for testing and outlining shallow placer deposits and for investigating material underlying gossan or capping. Their usefulness is limited to shallow-lying deposits, and the cross-sectional area of the opening is kept as small as possible. Although this sort of exploration has been almost completely replaced by the so-called placer-type of churn drill, in areas of low wages and excess labor, pits may still be used. Or, an individual may sink these shallow shafts on his own and disregard any assignment of wages to the project. The main difficulty encountered when sinking pits--apart from sampling difficulties--is the likelihood of the walls caving. An easy and economical way to support the opening is to sink a pit with a circular shape--say 3 to 5 ft. in diameter--depending on the depth. About every 3 to 4 ft. of advance, place a circular iron rim (wagon wheel tire) in the opening. Between the tire and the soil or gravel a few pieces of 2- x 4- or 2- x 6-inch lagging are placed (one-inch planks may well be sufficient). Ordinarily, less than 50 percent of the periphery will require support; that is, tight lagging is not necessary.

Generally all of the material removed is considered as the sample. The volume of the pit from which it came is carefully measured. If an alluvial deposit is under investigation, the sample is run through a short sluice or treated with a rocker. If water is encountered and causes trouble, a small, gasoline operated pump may remove it, but the discharge from the pump should pass through the sluice. For gold, final evaluation is in cents per cubic yard. In testing other than alluvial deposits, the material removed from the pit must be mixed and sampled as heretofore described.

H. Churn drilling may be used. Shallow, small-diameter holes are drilled to investigate alluvial deposits. During more advanced stages of exploration, the churn drill has been widely used for investigating and sampling flat-lying or massive deposits like coal beds, limestone beds, and massive, low-grade copper bodies as deep as several thousand feet. In recent years variously designed rotaries and downhole percussion drills have been replacing the churn drill.

I. Diamond drill may be used. Nearly every type of rock has been investigated by diamond drilling. Suitable for drilling deposits of any attitude (vertical to flat), this drill is most suitable under conditions giving at least 50 percent core recovery up to the hardest rock. Holes may be drilled from shallow to great depth and at any angle. The modern machine with its better designed core barrel has increased the core recovery.

Whatever method is used, it is absolutely necessary to remove all of the cuttings from each section of the test shafts, pits, and drill holes. This recovery is difficult with heavy minerals, but if it is not accomplished, contamination (salting) of all the following sections of the hole will result and give misleading information regarding their value and position. Assigning the wrong location of the values will cause an interpretation that practically guarantees failure of the venture. Recovery of both core and sludge from diamond drilling may be required to get a fair average for some materials.

The foregoing outline briefly surveys customary methods used in prospecting. Large companies may make use of more scientific methods, or spend great sums for drilling, tunneling, and shaft sinking. A discussion of this advanced prospecting (now called exploration) is beyond our purpose.

blank

DEVELOPMENT

LOCATION OF OUTCROPS BEFORE STARTING DEVELOPMENT

During the beginning stages of the prospect, the terms <u>prospecting</u> and <u>development</u> were used with a similar, if not identical, meaning. Once it becomes evident that a constant and definite flow of ore can be maintained, the term prospecting drops out of the picture. From then on a development program is established. Sufficient to say that this program will include (1) finding new ore (still really prospecting); (2) converting probable ore to positive ore by accomplishing the necessary development work; and (3) bringing the possible or prospective indications to the probable classification. Immediately following discovery, the small operator is concerned to convert evidence that supports item (3) to a more positive indicator that will show sufficient merit to finance or favorably dispose of the prospect.

As has been explained, the prospector should confine his work to the vein. A minimum of sinking, tunneling, or other excavating projects should be planned outside of the vein. Funds are limited and every possible dollar should further expose ore and indicative geological evidence. Supporting geological information may possibly be obtained without including the vein in the program but the cost and speculative nature is usually beyond the resources of the small operator.

Figure 3 emphasizes several points of interest. If a topographic map is not available,* a little careful sketching, estimating, and close guessing will provide reasonable detail. Strikes and dips ordinarily do not maintain uniform directions for any great distance. Also, the length of a claim can not exceed 1,500 ft. Figure 3 covers about 1,000 ft. along the strike. Using relative (assumed) elevations and walking out the contours with the help of either a Brunton compass or a hand level, or both, a map may be made to show the major topographic changes. On the resulting sketch may be plotted the outcrop with its various irregularities resulting from the topography. In the figure, the vertical dip or 90° position is indicated by the dotted line. Note that this line is perfectly free of irregularities even when crossing two stream channels. The dashed outline gives the course of a vein dipping at 60° <u>toward</u> or into the hillside. Note that the line curves <u>upstream</u> at each end where the vein crosses the two gulleys. And finally, the full line suggests the position for a vein dipping 60° <u>away</u> from the hill. Here the crossing of the depressions resulting from the stream action curves the exposure down stream. For several reasons, the position shown in Figure 3 may not truly represent the surface exposure: dips and strikes do not ordinarily maintain uniform values for distances exceeding a few hundred feet; an accurate assumption of values for plotting the data may be difficult; and loose surface material may be thick enough to cause an error of many feet in the apparent intersection of the vein and the surface. This latter position departs increasingly from the true position as the dip decreases and the surface flattens out.

---

*Sources: U. S. Geological Survey; Forest Service; Bureaus of Mines and Geology of the different states.

In general, prospecting work should be planned so that it can be driven at right angles (perpendicular) to the true strike. Only then will the maximum depth be intersected for a given distance. This procedure will also result in minimum lengths of tunnels for maximum exposure of the vein along the dip. The prospector is cautioned against attempting an excessive vertical exposure. Until the commercial character of the deposit is clearly probable, long tunnels to intersect great depths should be avoided: about 100 ft. vertically should be the limit. This limit can be considerably greater if the investigation is by diamond drilling. If the slope of the hillside is relatively flat and the vein does not dip steeply (less than about 50°), the relative costs of tunneling and sinking for the desired depth should be determined. A given depth of exposure may be less costly and quicker by sinking.

Points for plotting the apparent position of the outcrop are not difficult to establish. First, a value for the true dip and strike should be decided upon. Two points on the outcrop, having equal elevations, will contain the strike. The dip is measured at right angles to the line connecting these points. In Figure 3, the line XYZ is constructed to pass through the outcrop at Y; at this point there is sufficient exposure to determine the dip and strike. The intersection of XYZ with the outcrop occurs at elevation 6,350. On XYZ are laid off points representing elevations 6,400, 6,350, 6,300, 6,250, and 6,200 which are to be used in plotting the outcrop for a vein dipping 60° into the hillside. On the opposite side will be noted elevations 6,450, 6,400, 6,350, 6,300, 6,250, 6,200, 6,150, and 6,100. These points are used to plot the 60° dip away from the hillside. The location of points representing these elevations on XYZ is determined by using the contour interval (in Figure 3, 50 ft.), and the dip, 60°. The constant interval between the elevation points on XYZ is conveniently found by graphic methods, as demonstrated in the upper left hand portion of the figure. By trigonometry, the interval is equal to 50 ft. divided by the tangent of 60°, or 28.9 or 29 ft. (That is, the interval equals the contour interval divided by the tangent of the dip). Either of these methods may be used.

Each underground contour line is 29 ft. horizontally from the preceding one and is shown by a continuous straight line through the respective elevation. Lines through each elevation parallel to the strike are extended until they intersect the corresponding surface contour lines. For example, if we refer to the elevations on the right of XYZ, a line drawn through 6,300 and parallel to the strike intersects surface contour 6,300 at two points on the right side and also at two points on the left side. Similarly, the remaining elevations are extended to establish points on the corresponding contour lines. When the points thus established are connected, the resulting line shows the probable position of the outcrop.

Three positions are shown in the figure. Each has the same strike and passes through a common location. This positioning was done to illustrate the relative position of a vertical vein, a vein dipping into the hillside, and one dipping in the direction of the ground slope. Selection of map scale and contour interval has been somewhat exaggerated to emphasize the irregularities.

The following conclusions may be drawn after one inspects the proposed tunnel site AB.

1. The direction is perpendicular to the strike.

2. The portal of the tunnel has an elevation of 6,200 ft.

3. If the vein dips away from the hill, the tunnel will penetrate it at C, about 60 ft. in.

4. The vertical height of the vein above C is from 6,320-6,200 feet or about 120 feet. This height is equal to 120 ÷ sine 60°, or 133 ft. along the dip. The intersection at C is found by extending the 6,200-foot elevation from XZ. (The elevation line equal to the portal elevation is always the one extended).

5. The vertical vein will be cut at D, about 150 ft. from the start.

6. There is between about 6,330-6,200 feet, or 130 ft. of vein above D.

7. In the final instance where the dip is into the hill, the intersection is at B, about 240 ft. from the portal.

8. For tunnel AB, the vertical height above B is between about 6,350-6,200 feet or 140 ft. And along the vein this height becomes 140 ÷ sine 60°, or about 162 ft.

9. Points C, D, and B occur where the elevation line on XYZ, equal to the portal elevation and intersects the tunnel's center line. In the immediate example these points are elevations for surface contour 6,200 and underground line 6,200.

10. Any direction other than AB will require a longer tunnel to expose a similar height of vein.

11. If a tunnel were started at E to explore the vein, it should be driven in the direction EC (which is, of course, the strike). Because of irregularities usually occurring in the strike, some swing to right or left might be expected to keep the heading in the vein. But in general, the tunnel should be directed along EC.

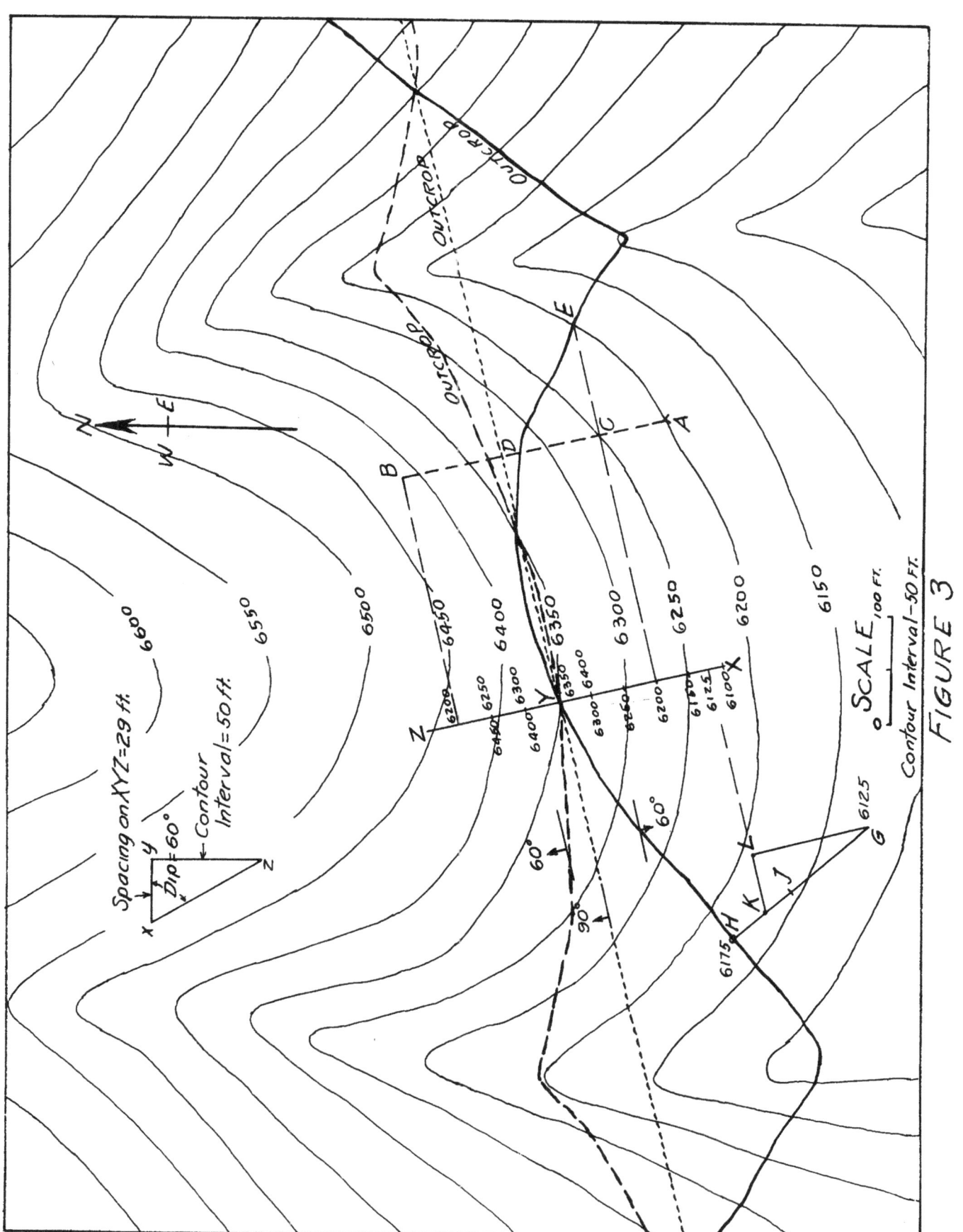

FIGURE 3

LOCATION OF SURFACE DEVELOPMENT OPENINGS

IMPORTANCE OF CORRECT LOCATION

The operator with limited capital must approach his development problems with a different objective in mind from that which would influence the large, well-financed and organized company. In general, very limited resources for the operation should restrict the development headings to the vein. Such restricted development would be followed until a desired minimum quantity of ore had been exposed. From then on, shafts, drifts, raises, etc., would most likely be driven in the country rock, were that choice available. The small operator must necessarily get the greatest amount of ore exposed for the least expenditure of time and money, which he can usually do by sinking or driving in the vein even though the vein may be very irregular in dip and strike. By following such a program, he will not only discover ore reserves with a minimum outlay for sinking and drifting, but the material thus removed may itself help pay the operating costs.

An irregular or sinuous drift is not too serious an obstacle to future use for haulage or mining. On the other hand, a shaft which follows the variation of dip presents several problems for efficient operation (higher maintenance, slower hoisting, somewhat greater length, and possibly more difficult support to mention only a few.) In spite of these inherent drawbacks, sinking should follow the vein. Initially, interest is in the present outlook, and future operating economies can temporarily be disregarded.

The shaft may be placed in one of several locations: in the hanging wall, in the vein, in the footwall; or it may start in the hanging wall, pass through the vein, and continue in the footwall. Regardless of the many lengthy discussions against the hanging wall position, the one chosen should represent the least outlay and bring forth the quickest information. Remember, the objective is to expose ore.

Before locating and assigning a direction to a prospect tunnel (known also as an adit--a term seldom used today), or to the dip and strike to an inclined shaft, the true dip and strike should be found. In many instances, such information is not difficult to discover. Occasionally, however, some uncertainty will arise. Then the problem is usually to find the true dip from apparent information. This solution is particularly important when diamond drilling to intersect veins at depth. The following example will explain this situation. Two points on the outcrop, some distance apart, have equal elevations. The strike or direction of the connecting line is the true one because both points have the same elevation. At or near one of the points, the true dip is desired. The apparent dip and strike is easily obtained at this location but considerable excavating would be necessary to expose additional information for measuring the exact dip. The true strike, N 40° W, was measured at a somewhat distant point. At the location in question, the strike is recorded as N 70° W, and the dip is 50° to the southeast. What is the true dip?

The solution may be made by simple trigonometry or by applying a simple graphic method. For the present purpose, the trigonometric formula is convenient.

$$\tan x = \frac{\tan y}{\sin z}$$

Where,

  x = true dip.

  y = apparent dip.

  z = angle between apparent strike and true strike.

  z = N 70° W − N 40° W = 30°

  y = 50°

  $\tan x = \dfrac{\tan 50°}{\sin 30°} = \dfrac{1.19175}{0.5000} = 2.38350$

  x = 67 1/2°

The value of the true dip is always greater than that of the apparent dip. If any two members of the formula are known, the remaining one may be calculated. If a shaft is to be sunk at this point, the dip should be about 67 1/2° if it is to stay in the vein; or, if a tunnel is to be driven to intersect the vein, its direction and length should be based on the dip of 67 1/2°. Knowing this angle would be of particular value for planning the tunnel, because considerable extra footage to reach the vein would be saved. Conceivably the plans for development could culminate in an unfavorable decision because of the apparent additional length mistakenly calculated on the basis of a 50° dip instead of the correct 67 1/2° dip if the dip is away from the portal. On the other hand, if the dip is toward the portal, and the wrong dip is used, the intersection would not be made when expected. Because of this error, discouragement or insufficient funds might terminate the work before the intersection was reached.

An additional illustration explains the effect of apparent dip. Suppose in Figure 3 a limited exposure is found at H. And suppose the strike of this exposure was mistakenly recorded as N 50 1/2° E and the dip 37° to the southeast. Not realizing that these measurements represent the apparent dip and strike, the prospector plans his tunnel along the line GH perpendicular to the exposure. Assume the tunnel is started at G with the elevation of the portal at 6,125 ft. The surface elevation of H is 6,175 ft. A simple calculation will indicate the point at which the tunnel presumably will intersect the vein.

If x is this distance (HJ) measured from H, then

$$x = \frac{(6,175 - 6,125)}{\tan 37°} = 66 \text{ ft.}$$

But because this information is based on the apparent dip, the tunnel would lack about 30 ft. of reaching the vein at K (because the vein actually dips 60°). In addition, only 50 ft. of vein along the dip is exposed. A tunnel started at G should have been driven toward L, a direction that is perpendicular to the true strike of the vein. Intersection would be at L and the height of vein exposed would be about $(6,235 - 6,125) \div \sin 60°$ or 127 ft. The distance GL would be about 13 ft. less than the tunnel along GH and more than twice the depth of ore would be exposed.

Before planning a diamond drilling program, a most careful study should be undertaken to determine true dips and strikes and surface positions of the vein. A saving in time and footage will result. Worth considering is the use of a diamond drill for prospecting instead of the customary shafts and tunnels. If the drilling campaign is carefully planned (and the services of an experienced mining engineer would be of great value at this point), the results from drilling can be as certain as cutting the vein with a tunnel. With few exceptions, the cost per unit of exposed vein will be less. But driving a tunnel or sinking a shaft does have the advantage of smaller immediate expenditures. In addition, openings are at once available for extracting ore.

By carefully planning the location of drill holes, footage often may be saved by deflecting a hole. Several hundred feet may be used several times (from the collar of the hole to the point of deflection). For an excellent discussion of the numerous applications of diamond drilling Cumming's book should be consulted (Cumming, 1951).

blank

TREATMENT OF ORE

METHODS OF TREATMENT

Most small operators are struck early with the desire to construct a concentrating plant. Little thought is given to the amount of ore that must be put in sight each day for even a very small operation. A few minutes time with pencil and paper should prove quite a shock to the ambitious builder of a mill. Certainly such a program should never be undertaken without the advice of both mining engineer and a competent mineral dressing engineer. This latter expert will have to make numerous metallurgical tests before he can design a suitable flowsheet. At least six months ore should be developed and reasonably good evidence be at hand for additional ore before a mill is considered.

There are many kinds of ores--simple and complex: sulfide ores, oxidized ores (carbonates and silicates and sulfates), and mixtures. It naturally follows that there will likely be just as many--perhaps more--methods for separating and concentrating them. Some of the gravity processes, extensively used in the past are seldom found in large plants today. But the application of stamps, tables, jigs, and amalgamation plates to many small-scale operations is still justifiable. The adoption of flotation, sink-float, cyanidation, electrostatic, and similar methods requires the services of an experienced ore dressing engineer. The investigation, design, and construction of a mill will closely approach a minimum of $1,000 per ton per day of output. An expenditure this large is better applied to proving up ore reserves. Of equal importance (but, sad to say, often neglected) is an adequate water supply.

WATER REQUIREMENTS

Water requirements for mills vary widely (from 2 to 20 tons of water per ton of ore). For flotation plants, 3 to 5 tons of water must be in the circuit for each ton of ore (Taggart, 1945, sec. 20. p. 12); cyanidation plants require about half again as much. Jigs or tables will take about the same quantity as does a cyanidation plant. In addition to the water in the circuit, additional water must be provided each day to make up losses (concentrates, tailing, evaporation). This additional amount ranges from a few hundred tons to several thousand tons per 24 hours.

A mill usually means a camp which, in turn, calls for an adequate domestic water supply. To meet these requirements, 100 to 200 gallons per day per capita must be available.

PRELIMINARY INVESTIGATION FOR CONCENTRATING PLANT

Assume that a 50-ton per day mill is under consideration (any size much smaller would represent too great a construction cost per ton). A concentrating program is most economically carried on with a minimum number of shutdowns. The justifiable assumption that the operation will close down only on week ends gives an operating time of 24 hours per day, 5 days per week.

The mine usually will be required to operate only on the day shift so far as breaking ore is concerned. Two shifts might be considered; but certain costs, mainly supervision and power demand, generally can be reduced by confining mining to one shift.

At a figure of 11 cu. ft. per ton for the ore, 550 cu. ft. per day must be broken. But this figure represents only the ore. Few mines are fortunate enough to require no waste removal to get the ore. This removal is ordinarily known as development work. Depending on management's policy of cost accounting, the development charges may or may not include exploratory work for additional ore. Regardless of how the costs may be apportioned, the waste rock from exploration must be disposed of and an estimate for this should be included in computing costs. At this point, it is not the cost that is of interest but the total material to be broken to maintain 50 tons of ore per 24 hours for the mill. Too many factors influence the amount of development required per ton of ore mined to undertake their discussion here. The size of the vein, its continuity, and the mining method will likely be most influential during the early life. For relatively shallow mining, some form of semi-open stope mining would prevail. An inspection of the literature indicates that about 20 to 30 percent of the cost will be charged to development (Vanderberg, 1932). This estimate will be assumed as applying to the volume of rock broken. If we use 30 percent as the ratio of development, the total minimum rock and ore to be broken each shift will be 786 cu. ft. This amount is equivalent to extending a 5- by 5-ft. drift about 22 ft. each day. At first glance, this development may not seem to be a great deal; but in one month it amounts to 480 ft.; and in one year, about 5,700 ft.

A rule that might well be considered before contemplating a concentrating plant is to delay the final decision until ore blocked out is of sufficient value to return at least the construction cost of the mill along with the mining, milling and smelting costs plus interest charges. A few successful exceptions to this policy are well known; but many more abandoned concentrating plants may be found because such a policy was not followed.

By far the best plan is to block out ore and then interest the financially able organization to take over. If the owner has the services of a competent mining consultant and legal advisor, there is little chance of his not getting fair treatment (Ricketts, 1943, Appendix on Forms and Agreements). Most of the forms given in Ricketts have been tested in the courts and deserve consideration when planning leases, options, etc.

HAND SORTING

Hand sorting, cobbing, or picking is the process of separating the mine-run material by hand (Taggart, 1945, sec. 19, p. 203; Peele, 1941, sec. 28, p. 15). Before undertaking such a program, many factors should be considered and studied carefully. This operation deserves investigation by the small producer. Among the numerous factors affecting sorting may be mentioned:

(1) Wages and labor; (2) degree to which waste and values are or may be separated; (3) distinguishing color or other physical characteristics of the ore especially under different types of illumination; (4) relative value of each product; (5) tonnage which must be sorted; (6) increase of shipping costs when grade is increased; (7) change in treatment charges; (8) more careful mining (blasting); (9) erection of facilities for sorting; (10) reduction of large pieces by hammer or power-driven crusher; (11) washing of mine-run ore before sorting; (12) effect on fines of washing; (13) removal of either discard or shipping material; and finally, (14) tons output and cost per shift.

Assigning estimates to the foregoing items is difficult. Too many factors depend on specific and local conditions but a few limited and general comments may be made.

(1) Wages and labor--men and women no longer able to perform hard, continuous work are usually employed. Boys and girls make very good sorters. Wages, therefore, will likely be much less than for skilled workmen, and will probably approximate a legal minimum wage.

(2) Data available indicates that from 0.5 percent to more than 30 percent of the untreated material may be separated.

(3) Daylight is best, but certain types of lighting which bring out the color of key minerals may be used.

(4) Calculations based on costs can usually indicate maximum values that can remain in discard and minimum value to which shipping material must be raised.

(5) Tonnage depends on the gross cost of operation, or the gross cost less that part of the cost derived from other sources.

(6) Many common carriers base freight charges on the value of the shipment with little regard for the weight.

(7) Treatment charges (mill or smelter), may increase or decrease with a change in grade (an increase could result if the sorting process increased above the minimum, material entailing penalties; and conversely, penalizable material may decrease by sorting.

(8) The quality of mining needs little comment: brittle, easily shattered ore may be broken too fine for good sorting.

(9) Material should be moved at maximum sorting velocity on a belt conveyor or revolving table with nearby bins into which one of the products is tossed. The product that stays on the belt is usually that in the greater amount. Even a generalized discussion of this subject rapidly becomes a major design problem. In place of a belt, a sorting floor, platform, bench, or revolving table may be used. Or, sorting at the chute as ore is withdrawn may be practiced. Some hand breaking may be necessary.

(10) If the mine-run grizzly is set too coarse, a crusher or hand reduction is necessary to unlock mineral and waste. It may be desirable to screen out mine-run fines.

(11 and 12) If color of products is useful to distinguish between them, washing may be required. Practical experience indicates that oversize from a grizzly is best suited to hand sorting. Mine-run or crusher product may require trommeling and washing.

(13) Of the two products, the one whose physical attributes can most quickly be recognized, and the one whose size makes for easy handling will ordinarily be removed.

(14) According to Peele (2nd Edition, p. 1651: these data are not in the 3rd edition), the theoretical data in Table 4 are suggested.

Table 4--Estimated (theoretical) Cost of Hand Picking Galena

Product	Wt. of single lump of galena, lb.	Time to pick one lump, sec.	Wt. galena picked in 10 hr., lb.	Cost per ton picked out, wages $1 per 10 hr.
6-in. cube	58.46	42	50,100	$0.040
5 " "	33.83	24	50,750	0.039
4 " "	17.321	12	51,970	0.038
3 " "	7.308	5	52,620	0.038
2 " "	2.165	3	25,980	0.077
1 1/2 " "	0.9134	2	16,442	0.122
1 " "	0.27061	1	9,743	0.205
3/4 " "	0.11421	1	4,112	0.486
1/2 "	0.03383	1	1,218	1.642

Consider the 3- to 6-in. sizes: data for this size-range suggests it as a minimum for both quantity and cost. Additional data in Peele suggests removing the 6- to 12-in. sizes to increase the tons per man-hour. Between those sizes, the total picked ranges from a few hundred pounds to several tons per man-hour. Lacking precise information, one may estimate 0.5 to 1.5 tons per man-hour. The lower figure would apply to sorting of less than 3- to 6-in. rock.

Formulas are available for solving detailed and complex problems in sorting (Peele, 1941, sec. 28, p. 15; Taggart, 1945, sec. 19, p. 203). Such problems usually involve a direct smelting product with several of the sorted products going to a concentrator). Neither a discussion nor a listing of these formulas is needed by the small producer.

Example of hand sorting

An example will illustrate the features usually of interest to the small operator. To avoid unnecessary lengths and complexity, a very simple gold-quartz ore will be investigated. Hand-sorting costs for a lead-zinc-silver ore or another such ore would be calculated in a similar way.

A gold-quartz vein occupies about 40 percent of the minimum width required for mining. The vein, tightly "frozen" to the walls, does not break completely free of the waste. The color difference between vein and wall rock makes hand sorting quite easy and rapid. Some of the quartz adheres to the wall rock. A sample across the mining face and a preliminary investigation suggests the advisability of using hand sorting. Assay of products will run: mine-run ore -- $21.50 per ton; discard -- $1.50 per ton. It is estimated the ore can be sorted at the rate of 0.8 tons per man-hour; wages for sorting are $1.50 per hour; freight to smelter for mine-run ore is $5.66 per ton; smelting for either mine-run or sorted product is $10.50 per ton. Freight on selected ore must be determined after ore grade is determined. The preliminary investigation indicates that about 60 percent of the mined product may be discarded.

$$1\text{-ton mine-run @ \$21.50} = 21.50 \text{ ton-\$}$$

$$60\% \times 1\text{-ton} \times \$1.50 = 0.6 \times 1.50 = 0.90 \text{ "}$$

$$40\% \times 1\text{-ton} \times \text{?} = 0.4 \times \text{?} = 20.60 \text{ "}$$

value of sorted product = 20.60/0.4

$$= \$51.50 \text{ per ton}$$

Mine-run to produce 1-ton of sorted material = 1 ÷ 0.4

$$= 2.5 \text{ tons}$$

A comparison of the two proposals--direct shipping or sorting--disregarding mining costs, taxes, insurance, compensation, etc., would be as follows:

Direct Shipping

2 1/2 tons @ $5.66	=	$14.15	freight
2 1/2 tons @ 10.50	=	26.25	smelting
Total	=	$40.40	
2 1/2 tons @ $21.50	=	$53.75	value of ore shipped
Less freight and smelting	=	40.40	
Net	=	$13.35	gross profit
or	=	$ 5.34	per ton

Sorting

1 ton @ $8.10	=	$ 8.10 freight
1 ton @ 10.50	=	10.50 smelting
2 1/2 tons @ 0.8T/hr @ $1.50/hr.	=	
3 hours @ $1.50	=	4.50 sorting
Total		23.10
1 ton @ $51.50	=	51.50
Less cost	=	23.10
Net	=	$28.40 gross profit
Or	=	$11.36 per ton of mine-run ore.

A considerable advantage is indicated if the mine-run material is sorted. The difference of approximately $6/ton between the two, further suggests that some of the variable factors could vary through a somewhat wide range and sorting would still be worth investigating.

If the percentage of vein and waste used above is reversed, direct shipping will yield $9.08 and sorting $14.19 gross profit.

To illustrate an additional effect, suppose that the solid vein confined between the two walls changes to a series of veinlets or stringers separated by country rock.

Sorting now becomes difficult and only 25 percent can be removed as discard. Also an appreciable part of the values has been deposited in the intervening stringers of gangue. Because of this change, the value of the discard rises to $10 per ton. Under these adverse circumstances, the direct shipping of mine-run ore would yield $7.12 per ton gross as compared to $6.43 for the sorted product. Probably neither ore could be shipped, for costs which have not been included would exceed the small profit indicated. Under such conditions, development should continue until either a favorable sale of the property can be negotiated or sufficient ore put in sight to return the cost of a concentrating plant under the plan previously recommended.

blank

DRILLING AND BLASTING

APPLICATION TO SMALL OPERATIONS

With one exception, suggestions for drilling and blasting will be confined to machine drilling. Only under exceptional conditions would hand drilling be resorted to. A few hand-drilled shots may be used to open up an outcrop in a remote and inaccessible location. With four-wheel drive vehicles, parachute drops, and helicopters, locations difficult to reach are becoming rare. A little known, but available means of opening the vein beyond the surface exposure would be the use of shaped charges. (see FURTHER RECOMMENDED READING: Austin, 1959; Austin, 1960; Draper, Hill and Agnew, 1948; Huttl, 1946). These explosive devices have been used to drill rough holes for later loading with explosives for simply loosening up rock masses. The cost of using shaped charges is high compared to that of conventional drilling and blasting. But they have the advantage of not requiring drilling equipment, the number used will be small, and nothing has to be returned to camp. At present the prospector will have to make his own shaped charge (this process is explained in the references quoted). The most suitable and versatile drill for the small producer or prospector who has advanced to the tunneling or sinking stage is the airleg drill (also called the jackleg). With several supplementary parts, this all-purpose drill with detachable bits (either carbon steel or tungsten carbide inserts) may be used quite efficiently for drifting, raising, sinking, and stoping. It may be used underground or on the surface; and on a platform as a jumbo-type drill rig. The jackleg, distinctly a one-man drill, is a real boon to the small operator.

There are many drill hole arrangements or patterns that may be used to break rock. While searching for and adopting the correct one for large scale mining has economic importance, the prospector had better confine his attention to one pattern. Figure 4A illustrates a drill round which will prove suitable in nearly all types of ground. Additional holes may be added if breaking becomes difficult. And if over-breaking occurs with the number shown, the amount of powder or its strength may be reduced. For blasting this round, a medium velocity (10,000 to 12,000 feet per second) semi-gelatin-type, 40 percent dynamite is suggested. The cartridge size is usually 1 1/4 by 8 in. Number 6 caps with ordinary safety fuse may be used for detonating the explosive. To time and fire the round, ignitacord and connectors are recommended.

The cut holes (number 1) would be loaded to about 2/3 of their depth. Remaining holes would be loaded about one-half full, although the lifters (number 5) might require an extra cartridge. These charges would be the standard recommendation. Increasing or decreasing the number of sticks after several trials should give the right charge.

Experimenting with the "burn" cut is not recommended. A proper arrangement of unloaded holes; relative diameters of each; spacing between them, and explosive charge must be carefully determined for each rock-type. An error in any of these variables could easily result in no ground broken. Such experimentation is hardly worth the prospector's time.

When mining has progressed to the point of steady output, experimental investigations would be undertaken to develop the most economical blasting round and charge (Blaster's Handbook, any edition). Figure 4A, the four center holes are the cut holes: they are detonated first and simultaneously. Holes 2 and 3 may be timed for the order shown; or the order may be reversed; or all four may fire as number 2. The back holes are number 4: they may be timed to go closely together or in some prearranged order. The bottom holes or lifters, number 5, should always fire last. Note especially holes 2, 4, and 5: they are slightly inclined so that they bottom just outside the desired boundaries of the drift. Neglecting such an incline in drilling will gradually "shrink" the heading to a size less than planned on. Right-hand number 5 hole is inclined sufficiently to provide for the drainage ditch. Other lifters should project sufficiently into the floor to provide for keeping the drift on grade. A commonly desired grade for drifts is 0.5 to 0.75 percent (6 in. to 9 in. per 100 feet). A grade board with an ordinary carpenter's level is used for maintaining grade.

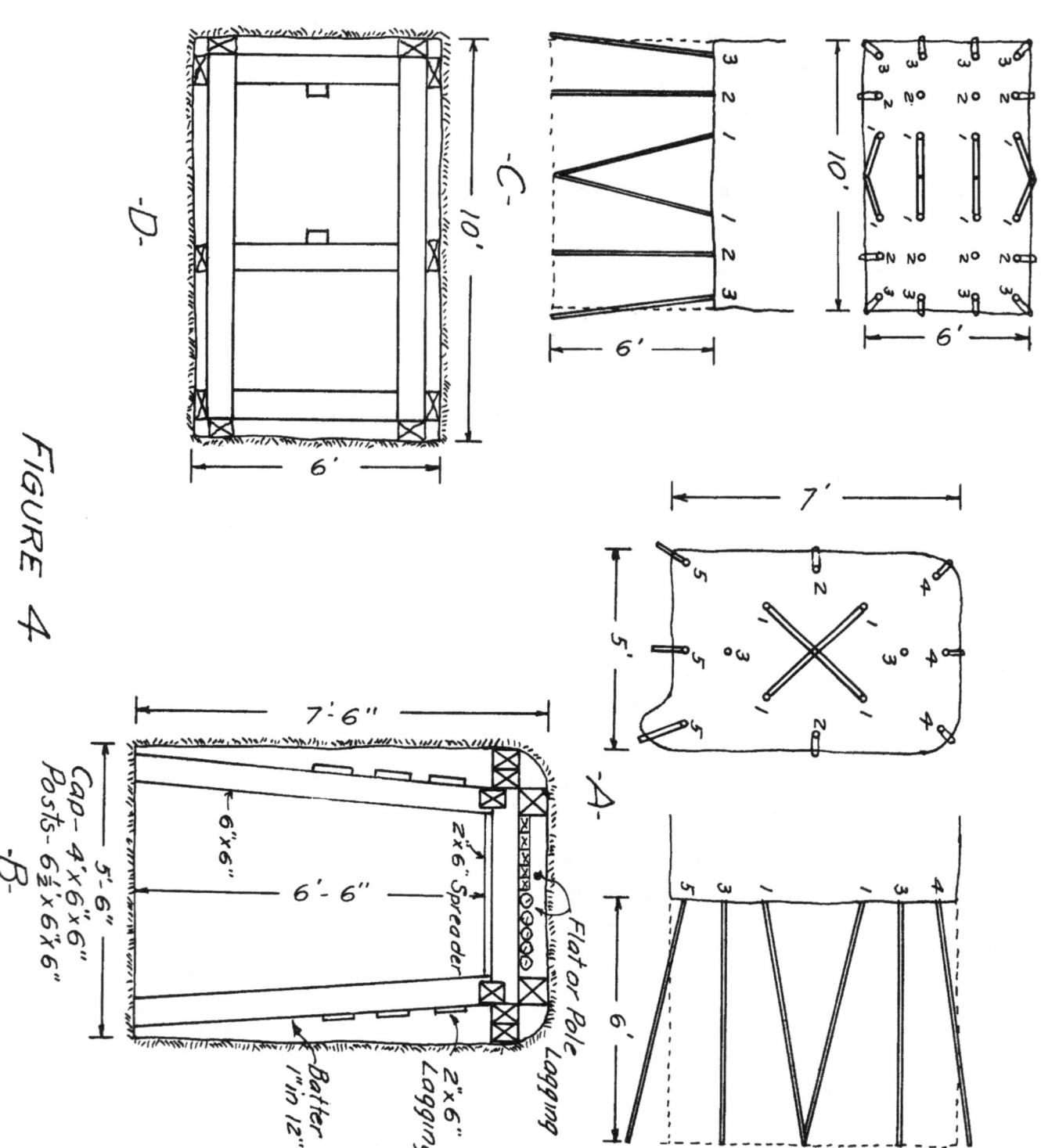

FIGURE 4

SUPPORT OF GROUND

METHODS APPLICABLE TO SMALL OPERATIONS

Figure 4B gives the detail for a simple drift set. Block as shown, at the ends of the cap and on top of the cap. The diagonal lines on the blocking indicate end grain in the blocking. It is very important that the blocking lies as shown; otherwise, the side pressure will buckle the cap and top pressure will cause the posts to fail. Fault zones and easily altered dikes may require larger timber than shown in the figure. Round timber should always have the bark removed.

For an additional means of support, the small operator should consider the use of rock bolts. The following literature will provide pertinent information.*

For additional suggestions about timbering, drifts, raises, or shafts, Peele should be consulted (Peele, 1941, sec. 10, p. 197-273).

Figure 4C gives the blasting arrangement for a two-compartment shaft. The numbers indicate the order of firing the holes. A shaft round usually requires a little more explosive per hole than does a drift round. Figure 4D shows the arrangement of the two compartments. The sets would be blocked at the corners and opposite the center divide. The amount of lagging would depend on the rock walls. Even though a bucket is used for hoisting material, guides and a safety crosshead should be used in the larger compartment. The smaller one is for the ladderway.

---

*Ohio Brass Co., Mansfield, Ohio. "Haulage Ways", Sept. 1956.
   Bethlehem Steel Co., Bethlehem, Pa. "Bethlehem Mine Roof and Rock Bolts".
   The Colorado Fuel and Iron Corp., Denver, Colo. "Mine Rock Bolts".
   Perfo Division, Sika Chemical Corporation, 35 Gregory Ave., Passaic, New Jersey. "Perfo Method for Roof-Bolting".
   These articles are only a few of the many sources of information.

blank

MINING METHODS

COSTS FOR VARIOUS METHODS

A detailed discussion of mining or stoping procedures is hardly necessary for the individual prospector developing a prospect. Considerable planning and development would have to be worked out and the numerous variations in the methods evaluated. If the prospect passes into the operating stage and if a substantial tonnage of positive ore becomes available, the choice of a mining method can then be decided upon. (FURTHER RECOMMENDED READING: Jackson and Gardner, 1936; Jackson and Hedges, 1939; U. S. Bur. Mines, Miners' Circ. 52, 1955; Peele, any edition.)

For supplementing these references, the old data in Table 5 may be of value (Wright, 1935). More up-to-date information indicates a considerable increase in tons per man-shift (Chandler, 1959, 1960): for room and pillar mining, 23.3 tons per man-shift; and for square-set mining, 3.9 per man-shift. This difference is the result of the high degree of mechanization today. Yet the small mine today will still more closely approach the ton output indicated by the earlier data. Even though there will be better mechanization than in the 1930's, the small producer still cannot approach that of the large mines. The application of loading machines and slushers will present no great problems in the small mines.

Table 5--Average Results for Different Mining Methods

(Compiled from table in Wright, 1935)

Mining Method	Man-hours per ton, all underground labor	Tons per man-shift all underground labor	Explosives, pounds per ton	Power kwh per ton
Open stopes	1.17	6.837	0.844	8.90
Shrinkage	1.33	6.015	1.656	15.33
Cut-and-fill	4.34	1.843	0.660	10.05
Square-set	5.15	1.553	1.363	34.9
Open cut	0.324	24.691	0.308	2.71
All underground mines	1.83	4.37	0.76	12.72

Tables 6 and 7 are from the article by Chandler (1959, 1960).

Table 6--Explosives Consumption for Various Mining Methods

Mining method	Ground conditions	Pillar mining	Stoping Range	Average
Square-setting	weak	0.20	0.30 - 1.08	0.5
Cut-and-fill	medium	0.30	0.50 - 1.23	0.7
Sub-level stoping	strong	0.26	0.33 - 0.59	0.4
Room and pillar	strong		0.67 - 1.14	0.8
Block caving	weak		0.08 - 0.19	0.14
Block caving	medium-strong		0.23 - 0.47	0.35
Open pit			0.10 - 0.53	0.28

(Explosives, lb. per ton broken)

Table 7--Timber Consumption for Various Mining Methods

Mining method	Board feet per ton mined Range	Average
Square-setting	12.0 - 19.7	15.1
Mitchell slice	7.5 - 10.5	9.3
Cut-and-fill	0.8 - 7.0	1.5
Shrinkage	0.4 - 3.9	1.9
Open stope (small)	0 - 1.7	0.7
Sub-level stoping	1.0 - 2.0	1.5
Block caving	0 - 2.0	1.2

Some of the methods in the tables will seldom apply in small mines whose depth normally will be above the 200-ft. level. Probably the higher figure in each case should be applied. Only a few of the smaller operations will equal the efficiency of the larger mines.

In conclusion, I direct the reader to Miners' Circular 52, (1955). This publication has a series of excellent illustrations explaining mining methods.

MINING COSTS

COST OF EQUIPMENT, SHIPMENT OF ORE AND LABOR

Accurate and detailed costs are necessary if economy is to be obtained and border-line properties kept in operation. The system adopted should produce cost-data for management at frequent intervals if management is to take maximum advantage of the information. Certainly a daily interval should be strived for. The information does not necessarily need to be reported in dollars. Consumption per ton; or advance per foot for the supplies used (caps, fuse, powder, timber, bits, drill steel, rails, etc.); or tons per man-shift or man-shifts per ton--any of these figures is apt to be more enlightening than dollars per ton. But while a well worked-out system along such lines is all very well for the established, operating mine, for the prospector or small mine owner developing the property for sale a similar procedure will not only be difficult to put into effect, but in fact may be of little use. Conceivably, too close attention to such detail could seriously handicap prospecting. The following tables on costs reflect essentially capital expenses, although there is a difference of opinion as to what charges should be credited to capital outlay.

General mining equipment

An itemized list of the numerous supplies required for even the most modest mining venture soon becomes extended. The items listed here will give those interested in financing a prospect a general idea of costs.

There are a large variety of makes, styles, and sizes for most items of mining equipment. To keep a list within reasonable bounds only a small number can be tabulated. Prices listed should be taken as a close guide, although they vary from place to place and undergo frequent changes. Table 8* lists the common items that sooner or later must be acquired.

Table 8---Mining Equipment and Supplies

Item	Cost new, dollars
Air Drills	
Jackleg, complete with feed leg, 36-in. or 48-in. feed, about 94 lb.	1025
Stoper, about 91 lb. automatic rotation	1100
Hose, 1-in. air with fittings, 50 ft.,	59
1/2-in. water with fittings, 50 ft.	37
Oiler	35

*Personal communication, R. B. Austin, The Coeur d'Alene Co., Wallace Idaho.

Bits, 4-point tungsten carbide inserts, 7/8-in. steel,
    1 3/8-in. dia.   12.10
    1 1/2-in. dia.   13.50

Drill Steel
    Jackleg, 7/8-in. Hex. collared and including bit connection

carbon steel,	2 feet	8.50 each
	4 "	10.40
	6 "	12.30
	8 "	14.20
	10 "	16.05
alloy steel,	2 feet	11.00 each
	4 "	13.25
	6 "	15.50
	8 "	17.75
	10 "	20.00

    Stoper, 7/8-in. QO, plain, including bit connection

carbon steel,	1 1/2 feet	6.85 each
	3 "	8.40
	4 1/2 "	10.00
	6 "	11.55
	7 1/2 "	13.15
alloy steel,	1 1/2 "	8.35 each
	3 "	10.25
	4 1/2 "	12.20
	6 "	14.10
	7 1/2 "	16.00

Bit Grinder, for carbide bits, air operated   145.00

Car, mine
    16 cu. ft. capacity   346.50
    20 " " "   396.00

Compressor*, 2-wheel mounting, 100 psi sea level rating
    125 cu. ft./min., gasoline engine   4715.00
    125 " "   Diesel engine   6365.00

Two or four wheels,
    250 cu. ft./min., gasoline engine   8560.00
    250 " "   Diesel engine   10,310.00

Four wheels,
    365 cu. ft./min., Diesel engine   13,820.00

---

*About $400-500 less if mounted on skids in place of wheels.

Table 8 (Cont'd.)

Hoist, with electric motor, capacity: 200-lb. cage, 1000-lb. car, 2000-lb. contents,	
40-hp. motor	7500.00
1000-1500 lb. capacity bucket, gasoline — electric motor	1515.00
Hoisting Cable, 3/4-in. dia., per 100 feet	42.40
1/2-in. dia. " " "	23.90
Pipe, per 100 feet	
2-in. dia. for compressed air	57.50
3/4-in. dia. for water	20.80
Pump, 100 gpm, 200-ft. head, motor-pump, 220/440 volt, 3-phase, 60 cycle, drip-proof motor	567 FOB
Track	
12-lb. rail per ton	249.00
16 " " " "	247.00
spikes per hundredweight	22.00
bolts " "	33.00
fish plates, per pair	0.69
Timber, ties, timber sets, framed per 1000 bf	75.00
Fan, ventilation	150-400.00
Ventilation pipe, 10-ft. lengths galvanized,	
8-in. dia. by 24 gage, per foot	1.10
11-" " " " " " "	1.15
15-" " " 22 " " "	1.50
Fittings for bends, 8-in. - 15-in.; 22 1/2° - 90°	5.15 - 21.85
Pick, 5 lb.	3.00
Shovel, round point	4.75
Saw, mine #17, 3 feet	6.50
Ax, 4-lb. single bit	5.50
Pipe wrench, 10-in.	2.75
Mucking machine, Eimco 20-35 cu. ft./min	3460.00

Most of the equipment listed in Table 8 may be bought second-hand and sold the same way. Depending on its condition and length of time used, second-hand equipment will cost or sell for 50 to 70 percent of the new price.

Diamond drilling equipment

Table 9 gives a minimum list of diamond drilling equipment for drilling to about 500 feet. Before the mine owner decides to purchase the equipment and undertake his own drilling a very careful cost analysis should be made. It is apt to prove exceedingly costly to train an unskilled man to operate a diamond drill.

Unless the footage is great and continuous drilling over several years anticipated, the small operator should contract his drilling. If a good diamond drill operator is available, however, owning the rig might prove economical.

A popular procedure with mining companies is to lease equipment. The contractor will supply a machine and 2-man crew, and furnish all supplies except bits, shells, and core barrels. The charge for underground drilling is $105 to $125 per shift; and for surface drilling $175 to $200 per day. The contract further permits the mine owner to cancel the contract on one day's notice if footage and/or core recovery is less than satisfactory.

In recent years the swivel-type core barrel permits a core recovery of 90 percent or better even in bad ground. The contract should specify the minimum recovery with stated penalties for failure to meet the contract. With a company-operated drill, better control over core recovery can be maintained through slower drilling. Contract drillers are, understandably, interested in making footage when the contract is on a straight price per foot.

In addition to the items in Table 9, tools, equipment, and supplies for fishing and supporting bad ground will sooner or later be required.

Table 9--Diamond Drill Equipment for 500-ft. EX Holes, Surface or Underground

Equipment	Costs, Dollars
450 ft. - 10-ft. EW rods @ $18.00	$ 810.00
50 ft. - 5-ft. " " @ 13.75	137.50
Water swivel, ball bearing	53.00
10-ft. swivel-type core barrel	80.00
Rod holding dog	29.15
Wrenches, grease, miscellaneous	100.00
	$ 1209.65
EX casing approximately $1.70 per foot	
Tripod for handling rods on surface,	170.00
Underground machine:	
Compressed air machine, complete with rod puller, mounting bar, accessories	2035.00
Water to be furnished by mine, minimum flow 4 gpm and 150 psi.	
Air to be furnished by mine, minimum 250 cfm at 90 psi.	
A good, used machine should be purchased for under	1500.00

Surface machine:
- Skid mounted, gasoline powered,
  - with screw-feed swivel head — 3100.00
  - with hydraulic swivel head — 3700.00
- Skid mounted pumping unit, air cooled gasoline engine, minimum of 17 gpm at 100 psi — 555.00
- 1 1/2-in. plastic pipe to bring water from supply point to drill site, 38 cents per foot
- Water tank to mount on truck for water haulage, 500 gal. capacity — 75.00

Used surface equipment: poor shape to good shape — 500-3000.00
  Pumping units about — 100.00

Summary:
- Underground, new — 3300.00
- Underground, used — 2200.00
- Surface, new — 4800-5500.00
- Surface, used — 3000.00

Bits and reaming shells
- EX coring bit, first grade diamonds — 89.00
- Reaming shell — 45.00
Diamond cost per foot drilled — 0.70-1.00

Average cost per foot:
(1) Company-owned equipment with competent driller
  (a) Underground, not including air or water — 3.75/foot
  (b) Surface — 4.50/foot
(2) Company-rented equipment with competent driller
  (a) Underground, not including air or water — 4.00/foot
  (b) Surface — 4.75/foot
(3) Contract*
  (a) Underground, not including air or water — 4.50/ft. min.
  (b) Surface — 5.50/ft. "

An itemized percentage breakdown of diamond drilling costs is:

Underground (not including air and water):

Labor	55 percent
Diamonds	28 "
Supplies and amortization, etc.	17 "

---

*Price here will depend on footage to be drilled. Contractor's price will drop with increased footage. On deep holes, contractor's price is very competitive with small mine operator because of cost of equipment for deep drilling.

Surface

    Labor   44 percent
    Diamonds   22   "
    Supplies and amortization, etc.   34   "

    With a company-owned machine, the cost approximates $50 per man per shift. Because a helper is necessary, the total becomes $100 per machine per shift. These costs, including air, water, hoisting, etc., are based on 20 feet drilled per shift. This estimate will vary from mine to mine.

    Unless business is quite slack, 1000 feet is probably the least footage an operator will contract for. Small contractors may accept smaller contracts.

Drilling contract

    The preceding data reflect costs in the Pacific Northwest. The range from minimum to maximum can be great but these data should provide a starting point for computing probable costs. When one considers diamond drilling, several very pertinent suggestions in Cumming (1951, p. 419) deserve to be mentioned. Among other things, the following items should be covered by the contract:*

  (1) Minimum and maximum depth of holes.
  (2) Extra cost for variation from above limits.
  (3) Minimum total footage.
  (4) Adjustment if drilling stopped by company before completion of contract.
  (5) Maximum moves between holes.
  (6) Amount of stand-piping done by contractor in any hole without extra charge.
  (7) Cost of casing hole.
  (8) Casing pipe left in hole by contractor
  (9) Size of core
(10) Wedging of holes
(11) Cementing holes.
(12) Surface or underground drilling.
(13) Length of pulls underground.
(14) Overburden payment basis.
(15) Sludge collection.
(16) Surveying of holes for bearing and inclination.
(17) Core boxes and handling to and from drill.

Type of contract (Sack, 1938, p. 46):
  (1) Straight footage basis, plus extras.
  (2) Cost-plus basis.
  (3) Rental basis, where one party owns the equipment and the other operates it.

---

* Core recovery should also be included.

An important item of cost mentioned by Cumming (1951, p. 421) is the percentage over and above the basic cost per foot. This extra cost (which will not appear in the contract estimate) includes items about which the contractor has little if anything to say. For example, in the preceding list many of the items, even though tied down by the contract, are indefinite; and in addition, company delays, company expense, engineering, etc., must be allowed for. If the contract footage is, say $10,000, the actual cost would be $10,000 plus (50% x 10,000) or $15,000. That is, experience shows that at least 50 percent of the contract price will be incurred because of extra expense and such a percentage should be provided for when estimating the total cost.

Road building

Very little prospecting (other than elementary, visual searching) will have been done before an access road becomes necessary. Road building requirements for mine access will range from a bare minimum to well-constructed roadways. Ordinarily, a bulldozer is sufficient for road construction. Including all costs, the services of a bulldozer may be rented for $10 to $12 per hour. Grades are usually quite steep. If the road is to be used throughout the year, gentler grades and crushed rock surfacing may become necessary. If the road is to be built on land under the jurisdiction of the Forest Service, the following requirements must be met.

If the road is to be built on forest service land, the district forest ranger must first be consulted. He will provide the prospector with information concerning restrictions, standards, engineering, maintenance, inspection, use-permit, protection of adjoining land, and other information. Maximum grades usually will not exceed 8 percent. Metal culverts are required at live stream crossings. Slash must be treated according to regulations.

As a guide toward building a mine road, the following information will be helpful.*

Construction costs would parallel those for the following examples.

Example 1

On simple road construction where there is little or no clearing, where common material is involved and side slopes (the slope of the hill) are 10 percent or less, and where minimum equipment and crew may therefore be involved, the cost may be as low as $500 per mile.

---

*Personal communication from E. F. Barry, U. S. Forest Service, Missoula, Montana.

Example 2

On a road with a moderate amount of blasting, the cost is again determined by character of material, side slopes, clearing and drainage costs. The construction time would be determined mainly by the size and effectiveness of equipment and crew. Here we assume a contour-type road, with balanced sections; that is, with material from cut side used to make fill on the lower side. For these two examples, the cost would be as follows:

(A) Width of road, 12 ft. without ditch. Average side slope, 25 percent; average cut slope, 3/4 to 1; no clearing.
Excavation per mile, 1800 cu. yd.

10% solid rock--180 cu. yd. @ $2.00	$ 360
90% common material--1620 cu. yd. @ $0.30	486
Total per mile	846

Culvert installations add to the $846.

(B) Width of road, 12 ft. without ditch. Average side slope, 50 percent; average cut slope, 3/4 to 1; medium clearing,

4 acres per mile @ $400	$1600
Excavation per mile, 6000 cu. yd., 25% solid rock--	
1500 cu. yd. @ $2.00	3000
75% common earth--4500 cu. yd. @ $0.30	1350
Total per mile	$5950

Culvert and drainage are additional costs.

Transportation rates for ores

Guides for the cost of shipping ores to a treatment plant may be found in Tables 10 (1), 11 (2), 12 (3), and 13 (4).

Truck freight rates, via B-Line, from mines in the vicinity of Metaline Falls, Washington, to Kellogg, Idaho, are as follows: (Also included are points within 5 miles of Kellogg). The minimum weight is 44,000 lb. (4).

Lead concentrate	$7.91 per ton.
Zinc concentrate	$6.93 per ton.

Table 13 gives rail freight rates from several points in Idaho and Washington to the Bunker Hill plant at Kellogg, Idaho.

The rates given in these tables should be checked with the shipping company agent. Larger quantities and regular shipments may result in more favorable rates.

Cost of explosives

Because the variety of explosives is so great, only those most generally used will be listed. For a general knowledge of their cost the prospector should consult Table 14 (5).

Table 15 lists types and costs of electric blasting caps (5).

For Idaho, the price of safety fuse in 1000-ft. rolls is: White or Black Sequoia, $14.40; Orange Sequoia or Dreadnought, $14.70; and Triple Tape, $16.85.

---

(1) Personal communication from R. F. Pettigrew, Union Pacific Railroad.
(2)   "   "   "   L. S. Davis, Northern Pacific Railroad.
(3)   "   "   "   Lundberg Truck Line, Mackay, Idaho.
(4)   "   "   "   A. Y. Bethune, Kellogg, Idaho.

Other information concerning the Bunker Hill Co. from the same source.

(5) Personal communication from S. M. Strohecker, Jr., Du Pont Company.

Table 10--Rates on Ore, South Idaho Points to Salt Lake City; Carload Lots.

Valuation, dollars per ton of 2000 lb., minimum 20 tons except as shown

Miles	From (Idaho)	10	15	20	30	40	50	60	70	80	90	100
281	Mackay	4.67	5.17	5.66	6.16	7.14	8.10	9.07	10.06	11.02	12.00	12.97
224	Montpelier	4.63*	--	5.13*	5.61*	6.08*	7.06*	8.02*	9.00*	9.97*	10.92*	11.90*
318	Victor	--	--	7.61	8.58	9.56	10.52	11.48	12.47	13.44	14.08	14.61
347	Ketchum	--	--	--	6.63*	7.61*	8.58*	9.56*	10.52*	11.48*	12.97*	14.08*
302	Buhl	5.17	5.66	6.16	6.63	7.61	8.58	9.56	10.52	11.48	12.47	13.44
272	Oakley	4.24#	5.17#	5.66#	6.16#	--	--	--	--	13.44	14.08	15.50
259	Declo	6.63	--	7.61	8.58	9.56	10.52	11.48	12.47	13.44	14.08	13.44
272	Oakley	5.17	--	5.66	6.16	7.61	8.58	9.56	10.52	11.48	12.47	13.44
	Valuation	125	150	175	200	225	250	300				
	Oakley	14.67	15.73	16.88	17.87	18.30	18.30	18.30				

*minimum--80,000 lb.
" 100,000 lb.

Table 11--Rates on Ores, North Idaho; Minimum Lots of 50,000 lb.

Distance, miles		Valuation, not exceeding $100/ton*	Valuation, $100 to $800/ton*
5 and under		5.30	6.70
10	and over 5	5.50	7.10
15	10	5.90	7.30
20	15	6.10	7.00
25	20	6.50	8.10
30	25	6.70	8.50
35	30	7.10	8.90
40	35	7.50	9.50
45	40	7.70	9.90
50	45	7.90	10.10
55	50	8.10	10.30
60	55	8.30	10.70
65	60	8.70	10.90
70	65	8.90	11.30
75	70	9.10	11.50
80	75	9.30	11.70
85	80	9.50	12.10
90	85	9.70	12.30
95	90	10.10	12.70
100	95	10.30	12.90
110	100	10.70	13.60
120	110	10.90	14.00
130	120	11.30	14.20
140	130	11.70	15.00
150	140	12.10	15.60
160	150	12.50	15.80
170	160	12.70	16.40
180	170	13.10	16.80
190	180	13.40	17.00
200	190	13.80	17.40

*2000-lb. ton

Table 12--Rates on Ore by Truck from Idaho Points to Garfield, Magma, Midvale, Murray, or Salt Lake City, Utah.*

Point of Origin

From	Cents per 100 lb.	From	Cents per 100 lb.
Alder Creek	74	Rob Roy Mine	90
Allison Mine	95	Sandy Creek	106
Bayhorse	80	Seafoam Mine	127
Blackbird Mine	78	Silver King	95
Black Pine Mine	76	Slate Creek	75
Buckskin Mine	100	South Butte Mine	85
Clayton Silver	80	Sunbeam Mine	90
Copper Basin	74	Baker, Idaho	100
Donahue Mine	85	Challis, Idaho	64
Fi Cappa**	74	Clayton, Idaho	80
4th of July Creek	85	Leadore, Idaho	100
Greyhound Mill	127	Mackay, Idaho	50
Grouse Creek	117	Salmon, Idaho	75
Hoodoo Mine	80	Sun Valley, Idaho	64
Leacocks Ranch	106	Tendoy, Idaho	100
Livingston Mill	90	Turtle Mine	75
McFadden Mine	117	Twin Apex (Salmon)	90
Montana Mine	106	Twin Peaks Mine	65
Parker Mtn. Mine	117	Washington Basin	106
Pope-Shenon	90	Wilbert Mine	53
Redbird Mine	80	Yankee Mine	90

*Rates to Tooele, Utah, are 10 cents/100 lb. more than those given. Rate from Clayton Silver Mine to East Helena, Mont., is $16/ton.
**A geological formation in this area is known as the Phi Kappa.

Table 13--Rail Freight Rates, Ores and Concentrates, to Bradley/Silver King, Idaho (1960)

Point of Origin

Value per ton	Clark Fork, Idaho Minimum 60,000 lb. (N. P. R. R.)	Metaline Falls, Wn. Minimum 100,000 lb* (C.M. St. P. & P.)	Northport, Wn. or Porthill, Idaho. Minimum 40,000 lb.
Under $30	$ 6.63	$ 6.16	$ 7.14
40	7.14	6.63	7.61
50	7.61	7.14	8.10
60	8.10	7.61	8.58
70	8.58	8.10	9.07

*If car of less capacity furnished, then 80,000 lb. minimum.

Table 13 (Cont'd.)

Over 70	---	$8.30**	---
80	9.07		9.56
90	9.56		10.06
100	10.06		10.52
Over 100	---		11.02

Milling Ore

Under 15	5.17
20	5.66
25	6.16

*This is maximum rate from Metaline Falls.

Table 14--Dynamite, 50-lb. Fiberboard Box, 1 1/4" x 8" Cartridges

Kind	Strength, percent	Cost per 50-lb. box*
Red Cross Extra	40	$26.00
"	50	26.50
"	60	27.00
Special Gelatin	40	28.70
"	50	29.45
"	60	30.20
Hi-velocity gelatin	40	30.45
"	50	32.20
"	60	33.80
Extra	--	27.00
Galex	--	27.75

*In 25 lb. fiber boxes add $0.50 per 100 lb.

Table 15--Regular Delay Electric Blasting Caps, Copper wire.
Price in Dollars per 100 caps

Delay	Length of wire, feet		
	6	8	10
Instantaneous	23.00	24.00	25.00
1st delay	30.50	31.50	32.50
2nd "	31.25	32.25	33.25
3rd "	32.00	33.00	34.00
4th "	32.75	33.75	34.75
5th "	33.50	34.50	35.50
6th "	34.25	35.25	36.25
7th "	35.00	36.00	37.00
8th "	35.75	36.75	37.75
9th "	36.50	37.50	38.50
10th "	37.25	38.25	39.25

The following items complete the listing of explosive materials: ignitacord or spittercord, $13.15 per 1000 ft.; connectors, $20.25 per 1000; Du Pont No. 50 blasting machine, $85; and #6 blasting caps, $32.75 per 1000.

Treatment charges

A detailed discussion or tabulation of treatment charges would become too involved for our purpose. Not only does each type of ore require a different schedule, but the numerous smelters will have different schedules. A sample, prepared as discussed under sampling, should be submitted to the smelter for an estimate of base charges, penalties, credits, deducts, and so forth. A better scheduling may be forthcoming if the miner can personally discuss his problems and ore with the smelter representative.

Before making a shipment, one should follow certain suggestions. Several of these, abstracted from the Bunker Hill general ore purchasing schedule, are reproduced here. If shipments are planned for other smelters, these same suggestions should be followed. A copy of the general clauses included in the schedules and an outline of the policy should be obtained from each smelter.

Weights and Samples--Settlements made on basis of information taken by the company....,

Representation--Presence of shipper or representative is urged and welcomed. This is especially true for the first shipment....,

Assays--In effect this paragraph states; the smelter, after sampling the shipment, returns to the shipper a portion of the sample; if the

shipper does not immediately return to the smelter the results
of his assay, the smelter assay shall prevail for settlements.
In case of disagreements of the assays an umpire assay will be made,
.....,

Hand Samples--Before actually making a shipment, the prospective
shipper must submit to the company, a small, representative sample of
two to three pounds. This sample will be used for assay and quotation
of schedule.

Small Shipments--All lots of less than 5 tons dry weight are subject to
an additional charge for sampling and assaying of $10 per lot.

Of particular interest to the prospector is the Siliceous and/or Basic Ore Schedule. The one which follows outlines the Bunker Hill schedule; but no doubt other smelters have a similar program. For the most beneficial schedule, the small operator is again reminded that a personal meeting with the smelter representative is to his best interest.

Siliceous and/or Basic Ore Schedule

Payments:
Gold: If 0.03 oz. or over, pay for 100% @ $30.50 per oz.
Silver: If 1.0 oz. or over, pay for 95% @ applicable quotation.
Lead: If 2.5% and under 25%, pay for 90% @ N.Y. quotation,
less 2.0¢ per pound.
Zinc: If 2.5% or over, pay for 50% @ 25% of East St. Louis quotation.

Quotations on day of receipt.

Deductions:
Base: $18.00 per short dry ton.
Flux: Charge lime under iron plus silica @ 8.0¢ per unit.
Silver: Charge for all silver paid for @ 8.0¢ per ounce.
Arsenic: Charge over 1.0% @ $1.00 per unit.
Bismuth: Charge over 0.1% of wet lead @ 50¢ per pound.
Moisture: Charge over 10% @ 20¢ per unit.
Sulfur: Charge over 16% @ 10¢ per unit.
Freight: Based on $27.17 per ton freight on lead to N.Y.
Any change in rate or tax thereon for seller's account.
Calculated on pounds of lead paid for.
Delivery: F. O. B. Smelter at Bradley, Idaho
Small Lots: Charge $10 per lot on all lots under 5.0 dry tons.
Samples: A representative sample of two or three pounds must
be submitted before any shipment will be accepted.
Freight Address: Bradley, Idaho

P. O. Address: Box 29, Kellogg, Idaho

Example--The application of the foregoing schedule is as follows:

Lead-Zinc-Silver-Copper-Gold Ore, Nov. 25, 1960

Moisture %	Gold, oz.	Silver, oz.	Copper, %	Lead, %	Iron, %	Insol. %	Lime %	Sulfur, %	Zinc, %	Antimony, %	Arsenic %
1.0	1.25	30.00	0.8	13.0	8.9	50.0	1.5	17.5	6.5	0.3	0.3

E and M J Quotations
   Silver: $0.91375 per ounce.
   Lead: 0.12 - 0.02 = $0.10 per pound.
   Zinc: $0.13 @ 25% = $0.0325 per pound.

	Per ton
Payments for Metals:	
Gold: 1.25 oz @ 100% @ $30.50 per oz.	$38.125
Silver: 30.0 oz. @ 95% @ $0.91375 per oz.	26.042
Lead: 13% @ 90% @ $0.10 per lb.	23.400
Zinc: 6.5% @ 50% @ $0.0325 per lb.	2.113
Gross value	$89.680
Smelting and Refining Deductions	Per ton
Base Charge:	$18.000
(8.9% iron plus 50% insol.) less 1.5% lime @ 8¢ per unit	4.592
Total smelting deductions	$22.592
Silver charge: (30 oz. @ 95%) @ 8.0¢ per oz.	2.280
Sulfur charge: (17.5% - 16.0%) @ 10¢ per unit	0.150
Total smelting and refining deductions	$25.022

Net value: $89.680 - $25.022 = $64.658 per ton

Settlement:
   1 ton dry weight @ $64.658 = $64.658

Example of lead concentrate--To present an idea of the procedure involved for a lead concentrate, a sample calculation of a purely imaginary ore is given. Data were submitted by the East Helena, Montana smelter.*

---

*Personal communication, S. M. Lane, East Helena, Montana.

Analysis of lead concentrate, November 1960

Gold	Silver	Lead	Copper	Zinc	Arsenic	Antimony	Bismuth	Moisture
0.10 oz.	75 oz.	55.0%	4.0%	6.0%	2.5%	1.0%	0.10%	10%

Values per ton

These values are calculated using the rates paid by the smelter for the various metals; rates are submitted in the appropriate schedule.

<u>Gold</u>: 100% -- 0.10 oz. @ $31.8125 = $ 3.18
<u>Silver</u>: 95% -- 75.0 oz. @ ($0.91375 less 1 cent) = 64.39
<u>Lead</u>: -- less 1.5%; 90% @ ($0.11 - $0.022)
     (55% - 1.5%) x 90% x $0.088 x 2000 lb. = 84.74
<u>Copper</u>:- less 1.0%; 3% x ($0.2935 - $0.09) x 2000# = 12.21

                                        Total paid for $164.52

Deductions per ton

Base of $10.50 is for 20% or less lead. Credit of 10 cents per ton is allowed for each percent lead over base percent.

(55% - 1.5% - 20%) = 33.5% receives credit.
33.5 x 10 cents = $3.35.

The East Helena smelter has a fluctuating wage adjustment to the base pay. For November this adjustment came to $0.95 per ton.

Base charge is: $10.50 + $0.95 - $3.35 = $8.10 per ton

Arsenic and antimony are combined; two percent is free with excess charged for at 50 cents per unit.

(2.50% + 1.00%) - 2.00% = 1.50% excess
1.50 @ $0.50 = $0.75

For bismuth, an amount equal to 0.1% of wet assay for lead content is free; excess is charged for at 50 cents per pound.

(55% x 2000 lb.) x 0.1% = 1.10 lb. bismuth free.
2000 lb. x 0.1% bismuth = 2.00 lb. bismuth in the ore.
2.00 - 1.10 = 0.9 lb. to be paid for.
0.9 x $.050 = $0.45
Total deductions per ton = $8.10 + 0.75 + 0.45 = $9.30
Value per ton = $164.52 - $9.30 = $155.22.
Freight per wet ton, Mackay, Idaho, to East Helena
    at a value of $165 per ton   = $14.75

Trucking, Clayton to Mackay @ $5 per ton	=	5.00
Bullion freight East Helena to NYC, increase over base	=	0.10
Total		$19.85
Net to shipper, per ton = $155.22 - $19.85	=	$135.37

Labor costs

In addition to hourly- or day's-pay schedules--usually established through negotiations with the union--several taxes are referred directly to the payroll. These taxes differ from state to state and even in the same state will depend on the incidence of unemployment, the hazard involved, and the amount of reserve in the state compensation fund. In Idaho, a prospective employer should correspond with the State Insurance Fund (Workmen's Compensation Law), State Capitol, Boise, Idaho, for the rates applicable to his business. These taxes, compensations, and insurance payments for Idaho all paid by the employer are approximately as follows:

Social Security--3% of first $4800 (the employer must also withhold the same amount from the employee's wages).
Industrial Accident Insurance and Occupational Hazard--1% to 18%*
Unemployment Tax--(3% state and 3% federal).

The application of these taxes would be made as follows:

Miner's base wage, $19.36 per day. If he works a 48-hr. week (Saturday will be 8 hr. at overtime pay-- 1 1/2 times base), his average daily wage becomes:

Overtime distribution	=	(19.36 x 1 1/2 - 19.36) ÷ 6 = $1.61/day
Base pay	=	19.36
Overtime pay	=	1.61
Payroll wage	=	$20.97

---

*Depends upon risk classification. In Idaho, the employer has the option between the State Insurance Fund and insuring with a private insurance company. The rates are usually similar. Early in 1961, private company rates were:(1) open-pit mines, $3.80 per $100 of payroll; and (2) underground mining, $6.86 per $100 of payroll. These rates vary with the total funds on deposit in the state fund. Rates with the State Insurance Fund for this period were: $3.04 per $100 of payroll for surface mines; and $5.49 per $100 of payroll for underground mines.

Payroll insurance and taxes:

Social Security--3% of $20.97	=	$ 0.63
Industrial Accident, underground mining, 6.86% of $20.97	=	1.44
Unemployment, 3% of $20.97	=	0.63
Total		$ 2.70

Total cost to employer = $20.97 + $2.70 = $23.67

Similar charges must be calculated for each employee.

Drifting or tunneling

The time or cost required to advance a drift may be estimated as follows: A 5-ft. wide by 7-ft. high drift is drilled with a jackleg machine. The hole pattern will be similar to that shown in Figure 4A and the holes are 6 ft. deep. Holes of this depth should break at least 5 ft. It will be assumed that 2 1/2 sq. ft. of face can be broken per hole. On the basis of this information, there will be 35 sq. ft. of drift, and 14 holes will be needed. The total footage of holes is 6 times 14, or 84 feet. This footage can be drilled at the rate of at least 40 ft. per hour with a drifter and certainly faster with a jackleg (Dickenson and Slager, 1960, p. 99). The drilling time will be two hours. Loading and blasting time will be computed at one hour.

If the drift is mucked out by hand, the rate expected should be two tons per man-hour (Dickenson and Slager, 1960, p. 68).* In addition to shovelling conditions, the tramming distance will influence the output. Taking the rock in place at 12 cu. ft. per ton, a 5-ft. advance for a 35-sq. ft. drift will be equivalent to about 15 tons. Using the above rate for hand shovelling, 3 3/4 hours are required to clean up the round. Total time per round:

Drilling	2 hours
Blasting	1
Mucking	3 3/4
Total	6 3/4 hours or one shift.

Advance per shift 5 feet.

This estimate allows little time for delays and ventilation. But the drilling and blasting time is liberal. With good workmen, the round in and round out per shift should be almost possible.

---

*Small working place, poor condition	1 1/2 to 2 tons/man-hour
Large working place, good condition	2 to 2 1/2 tons/man-hour
Off a smooth floor or slick sheet	2 1/2 to 3 tons/man-hour
In a shaft bottom	1/2 to 1 1/2 tons/man-hour

The above estimate assumes one man drilling and two shovelling out the round. During loading and blasting the extra man helps the miner; otherwise he timbers, lays track, and so on.

If a mucking machine had been available (20-35 cu. ft. per minute) the apparent time required would be about 8 minutes. To this time must be added the time and delays for moving cars. This extra time is difficult to estimate. If 20-cu.-ft. cars are used (and this size is quite common), 9 loads will be necessary. The farther the tunnel advances beyond the portal (or from the shaft station) the more time must be allotted to handling each car load. On well-maintained track, with the grade in favor of the load, a trammer should move about 100 ft. per min. The time to the load and return will be about 3 min. Approximately 4 to 5 min. per car will be consumed per 100 to 200 ft. With two cars available, a loading machine would require about one hour to clean out the round. In the preceding discussion a tunnel operation was assumed; no delays would be incurred waiting for the cage and each man would take turn-about loading and pushing out a car load. Unless the ground is exceptionally heavy and difficult to hold, timbering can lag several rounds behind drilling and can be placed during the drilling cycle. But if the heading is a drift advancing from a shaft station, an additional allowance must be made for hoisting unless standby cars or buckets are available. An additional assumption should be made: if the workmen are in no way interested in economical effort and have little interest in the ultimate outcome of the venture, a very decided lowering of units per man-hour must be expected. This simply means less than 5 feet advance per shift for the two men (or 2 1/2 feet per man-shift as previously estimated). The overall feet-per-man advance during a shift will be still less if additional outside help (foreman, blacksmith, etc.) is included.

Shaft sinking

Figure 4D illustrates a suitable shaft layout for prospecting, later development, and mining. An area much smaller than that posited will materially reduce the space needed for drilling, mucking, and timbering, as well as curtail later hoisting. Slowing down these operations will result in an increased cost per foot. A pure prospect shaft would be about 5 ft. by 5 ft. rock size. Figure 4C shows the blasting round for this shaft. At 2 1/2 sq. ft. of rock surface per hole, 24 holes are necessary. The holes, drilled 6 ft. deep, are assumed to break to 5 ft. The total footage required is 144 ft. per round.

Including delays, 6 in. per min. may be taken as a conservative drilling rate for jackhammer machines. The drilling time totals 4 hrs. and 48 min. With the shaft area of 60 sq. ft. and a depth of 5 ft. broken, the volume for each round is 300 cu. ft. This figure converts to 25 tons, if 12 cu. ft. per ton is used as a constant.

A shaft can be hand mucked at a rate of 1/2 to 1 1/2 tons per man-hour (Dickenson and Slager, 1960, p. 68). Assuming 1 ton/man-hr. (good conditions: little water and rock easy to shovel) for two men, and allowing one hour each for delays per shift, one can compute the cleaning-up time.

$$25 \div (2 \times 7) \text{ man-hr./shift} = 1.8 \text{ shifts}$$

Multiplied by the usual 8-hr. shift, this figure gives a time of 14 hr. 24 min.

The total time for putting in a shaft round is:

Drilling	4 hr. 48 min.
Loading and blasting	1 30
Mucking	14 24
Total	20 hr. 42 min.

Timbering, with two men for one shift, will add an additional 8 hr.

Total time = 20 hr. 42 min. + 8 hr. = 28 hr. 42 min.

This figure converts to 3.6 shifts.

The advance will be at the rate of 5 ft. $\div$ 3.6 = about 1.4 ft. per shift.

The foregoing computation estimates the labor directly connected with drilling, mucking, and timbering with two men per shift. To this number should be added at least one additional man. His duties would include operating the hoist and other surface duties incidental to the underground work. If utmost economy is to be practiced, the top-man could serve as foreman, hoistman, blacksmith, timekeeper, etc. Most likely, there would be a fourth man as foreman. He and the hoistman would share surface duties.

The estimated 1.4 ft. per shift advance would certainly be close to the minimum (Krumlauf, 1954, p. 12). A total of 144 ft. drilled for two men is conservative (72 ft. per man). Each should drill 90 to 100 ft. per shift (Dickenson and Slager, 1960, p. 74 and 124). This advance, 8 in. per min., would reduce the drilling time by about one-third. Under ordinary conditions at least an additional one-fourth of timbering time would be saved. And finally, 1 1/2 tons per man-hour for shovelling and loading in the shaft could also be expected, which would represent an additional one-third reduction. All in all, the favorable savings of about one-third would correspond to a 1.9-ft. advance per shift. Delays incidental to pumping water from a wet shaft would decrease the estimated footage in either case.

Supplies and other items usually represent 40 percent of the total cost, which means that labor's share is 60 percent. In figures that Chandler (1959, 1960) gives for several mining methods, the labor charge approaches 60 percent.

Correctly speaking, complete costs should include an apportionment of interest and expenditures for equipment. This apportionment is difficult to estimate unless the life of the equipment or of the mine is known and unless resale value is available.

Many of the more permanent pieces of mining equipment may be rented (Krumlauf, 1954, p. 14). In this event the rental, which should be included when estimating or determining costs, will add materially to the cost per foot.

Before undertaking to rent equipment, a firm effort should be made to decide the length of time required for the venture. It does not take too many months for rental to equal new or second hand costs credit with resale value. But it must not be overlooked that rental will require a much less immediate outlay. Lack of funds at the beginning may excuse sacrificing long-range economy.

The previously determined pay of the one miner ($23.67 per day) would represent 60 percent of the cost per shift. The total would be $23.67 ÷ 60 percent or $39.45 per shift for labor and supplies.

SERVICES OF THE IDAHO BUREAU OF MINES AND GEOLOGY

The Idaho Bureau of Mines and Geology in Moscow offers many services free to the residents of Idaho. And for a nominal sum the Bureau has mining, metallurgical, and geological publications for sale describing various areas and research in the state. A list of these publications may be had on application. Although many of the early, important bulletins and pamphlets are out of print, responsible persons may borrow these for a limited time.

Another service offered by the Bureau is the identification of minerals, rocks, or other natural materials.

Mineral dressing or treatment tests may be made if the results would benefit all the claim-holders in an area or all the prospectors for a certain commodity. Geological mapping and mineral investigations are done in the most promising sections of the state, both by state projects alone and in cooperation with the Geological Survey and U. S. Bureau of Mines. When funds and personnel are available, prospectors and others interested in mining in the state are visited during the summer season. Inquiries about any problem concerning the mineral industry are welcomed.

In those instances where an examination or consultation is required to prepare a property for exploration or sale, the consulting mining engineer or geologist must be called upon. Such projects are not functions of the State Bureau.

blank

SERVICES OF THE FEDERAL GOVERNMENT

In addition to services afforded by the U. S. Geological Survey and the U. S. Bureau of Mines, the Office of Mineral Exploration (OME), a subdivision in the Department of the Interior, has been established to give financial assistance to promising prospects.

The purpose and extent of this help may best be explained by quoting directly from an OME circular:

> Briefly stated, the Office of Minerals Exploration offers financial assistance to firms and individuals who would like to explore their properties or claims for one or more of 32 mineral commodities listed in the OME regulations. This help is offered to applicants who ordinarily would not undertake the exploration under present conditions or circumstances at their sole expense and who are unable to obtain funds from commercial sources on reasonable terms.

For more detailed instructions the prospector should ask OME for:

(1) OME Form 40, an application form.
(2) OME Booklet "Minerals Exploration Program".
(3) Regulations 30CFRChIII.

The suggested literature may be obtained from OME in Washington 25, D. C., or from one of the field offices. Region I embraces Alaska, Idaho, Montana, Oregon and Washington. The address is,

> OME,
> South 157 Howard Street,
> Spokane 4, Wash.

Under the foregoing program, technical assistance and financial aid may be made available to those who wish to search for:

Antimony	Mica (strategic)
Asbestos (strategic)	Molybdenum
Bauxite	Monazite
Beryl	Nickel
Cadmium	Platinum group metals
Chromite	Quartz crystal (piezo-electric)
Cobalt	Rare earths
Columbium	Rutile - brookite

Copper
Diamond (industrial)
Fluorspar
Graphite (crucible flakes)
Kyanite (strategic)
Lead
Manganese
Mercury

Selenium
Talc (block steatite)
Tantalum
Thorium
Tin
Uranium
Zinc

REFERENCES CITED

Bateman, A. M., 1950, Economic mineral deposits, 2nd ed.: John Wiley and Sons, Inc., New York, 915 p.

Billingsley, Paul and Locke, Augustus, 1941, Structure of ore districts in continental framework: Trans. Am. Inst. Mining, Metall. Engineers, Vol. 144, p. 9-64.

Chandler, J. W., 1959, 1960, Mine development and mine operating costs: Part II - Economic evaluation of proposed mining ventures: Mining Cong. Jour., Nov. and Dec. 1959 and Jan. 1960.

Cumming, J. D., 1951, Diamond drill handbook: J. K. Smit and Sons (Canada), Ltd., Toronto, 501 p.

Dickenson, E. H., and Slayer, T., Jr., 1960, Rock drill data: Ingersoll-Rand Co., New York, 399 p.

E. I. du Pont de Nemours, 1958, Blasters' handbook; E. I. du Pont de Nemours and Co., Wilmington 98, Del. (Any of the numerous editions all suitable).

Emmons, W. H., 1917, The enrichment of ore deposits: U. S. Geol. Survey Bull. 625, 530 p. (out-of-print).

_____, 1937, Gold deposits of the world: McGraw-Hill Book Co., Inc., New York, 562 p. (out-of-print).

Farrell, J. H., and Moses, A. J., 1912, Practical field geology, McGraw Hill Book Co., Inc., New York, 273 p. (out-of-print).

Golden Press, 1957, Rocks and minerals: Golden Press, Inc., New York, 122 p.

Hoover, H. C., 1909, Principles of mining: McGraw-Hill Book Co., Inc., New York, 199 p.

Hulin, C. D., Jan. 1945, Factors in the localization of mineral districts: Am. Inst. Mining Metall. Engineers, Mining Technology, T. P. 1762, 17 p.

Jackson, C. F., and Hedges, J. H., 1939, Metal mining practice: U. S. Bur. Mines Bull. 419, 512 p. (out-of-print).

Krumlauf, H. E., 1954, Exploration and development of small mines: Arizona Bur. of Mines Mineral Tech. Series No. 68, Bull. 164.

_____, (Editor), 1960, Surface mining practice, a symposium: College of Mines, Univ. of Arizona, 131 p.

Loomis, F. B., 1948, Field book of common rocks and minerals: Putnam and Sons, New York.

Mining Truth, Nov. 1, 1929, Cover illustration: Mining truth, Spokane, Wash. (This magazine is no longer printed).

Newhouse, W. H., (Editor), 1942, Ore deposits as related to structural features: Princeton Univ. Press, Princeton, New Jersey, 280 p.

Nininger, R. D., 1954, Minerals for atomic energy: D. Van Nostrand Co., Inc., New York, 367 p.

Parks, R. D., 1957, Examination and valuation of mineral property, 4th Ed.: Addison-Wesley Press, Inc., Cambridge, Mass. (Some earlier editions by Baxter and Parks), 515 p.

Peele, Robert, (Editor), 1941, Mining engineers' handbook; 3rd Ed.: John Wiley and Sons, Inc., New York, 45 sections.

Pierce, J. H., and Kennedy, T. F., 1960, Mine examination, reports, valuation: Pierce Management Corp., Scranton, Penna., 255 p.

Ricketts, A. H., 1943, American Mining Law: California Div. of Mines Bull. 123, 4th Ed., 1018 p.

Sack, Walter, June, 1938, When planning to diamond drill: Engineering and Mining Jour., p. 46.

Savage, C. N., 1961, Economic geology of central Idaho blacksand placers: Idaho Bur. Mines and Geology Bull. 17, 160 p.

Schwartz, G. M., 1939, Hydrothermal alteration of igneous rocks: Geol. Soc. America Bull. V. 50, p. 181.

Staley, W. W., 1948, Distribution of heavy alluvial minerals in Idaho: Idaho Bur Mines and Geology, Min. Res. Report No. 5, 12 p.

_____, 1960, Gold in Idaho: Idaho Bur. of Mines and Geology Pamph. 68, 2nd Ed., 53 p.

State Inspector of Mines, 1959, Mining laws of the state of Idaho: 120 p. (Obtain from State Mine Inspector, Boise, Idaho. 25 cents).

Taggart, A. F., 1945, Handbook of mineral dressing: John Wiley and Sons, Inc., New York, 22 sections.

Vanderbert, W. O., 1932, Factors governing the selection of the proper level interval in underground mines: U. S. Bur. Mines Inf. Cir. 6613, 18 p. (out-of-print).

Wright, C. W., 1935, Essentials in developing and financing a prospect into a mine: U. S. Bur. Mines Inf. Cir. 6839, 22 p.

FURTHER RECOMMENDED READING

The following additional reading is suggested as a guide if information of a more specialized nature is desirable. This list is not comprehensive, but most of the references themselves contain almost an exhaustive list of reference material.

Publication dates have been deliberately omitted from several of the references: Although many of these books occur in later editions, for background reading the early editions are about as informative as the latest; and the reader might pass up an older volume because of the difference in dates.

Because some of the material is marked out-of-print, a list of several second hand book dealers is included. Probably at least one of these dealers will have for sale most, if not all of the out-of-print publications.

U. S. Bur. Mines, 1955, Accidents from falls of rock or ore at metal and nonmetallic mines: Miners' Cir. 52, 85 p.

Austin, C. F., Dec. 1960, NMIMT conducts shaped-charge research: New Mexico Inst. of Min. and Tech., Alumni Bull., p. 26-30.

_____, 1959, Lined cavity shaped charges and their use in rock and earth materials: New Mexico Bur. Mines and Mineral Resources Bull. 69, 80 p.

Bureau of Land Management, Cir. No. 1941 on U. S. mining statutes.

Draper, H. C., Hill, J. F., and Agnew, W. G., 1948, Shaped charges applied to mining: Part I--Drilling holes for blasting, U. S. Bur. Mines Rept. Inv. 4371, 12 p.

Eaton, Lucien, 1934, Practical mine development and equipment: McGraw-Hill Book Co., Inc., New York, 405 p. (out-of-print).

Engineering and Mining Journal, New York

Farrell, J. N., and Moses, A. J., 1912, Practical field geology: McGraw-Hill Book Co., Inc., New York, 273 p. (out-of-print).

Gunther, C. Y., 1912, The examination of prospects: McGraw-Hill Book Co., Inc., New York, 222 p. (out-of-print).

Hoover, T. J., The economics of mining: Stanford University Press, 551 p.

Hunt, S. F., 1936, Mining geology outlined: (Published privately by the author, Salt Lake City, Utah), 129 p.

Huttl, J. B., 1946, The shaped charge: Engineering and Mining Jour., May 1946, p. 58-63.

Idaho Bureau of Mines and Geology: Moscow, Idaho, List of publications.

Jackson, C. F., and Gardner, E. D., 1936, Stoping methods and costs: U. S. Bur. Mines Bull. 390, 296 p. (out-of-print).

Jackson, C. F., and Hedges, J. H., 1939, Metal mining practice: U. S. Bur. Mines Bull. 419, 512 p. (out-of-print).

Lewis, R. S., Elements of mining: John Wiley and Sons, Inc., New York.

Lindgren, Waldemar, Mineral deposits: McGraw-Hill Book Co., Inc., New York.

McKinstry, H. E., 1948, Mining geology: Prentice-Hall, Inc., New York, 680 p.

Mining World, San Francisco.

Peele, Robert, (Editor), Mining engineers' handbook: John Wiley and Sons, Inc., New York (any of the three editions).

Prospecting for Uranium, 1949, Supt. of Documents, Washington 25, D. C., 123 p.

Spurr, J. E., 1926, Geology applied to mining: McGraw-Hill Book Co., Inc., New York, 361 p. (out-of-print).

Staley, W. W., 1944, Elementary methods of placer mining, Idaho Bur. Mines and Geology Pamph. 35, 13th Ed., 28 p.

_____, 1937, Design of small wooden headframes: U. S. Bur. Mines Inf. Cir. 6943, 37 p. (out-of-print).

_____, 1939, Introduction to mine surveying: Stanford University Press, 275 p.

_____, 1949, Mine plant design, 2nd Ed.: McGraw-Hill Book Co., Inc., New York, 540 p.

Stoces, Bohuslav, 1958, Introduction to mining: Vol. 1--Text, 710 p; Vol. II--Illustrations, 368 p.: Pergamon Press, Inc., New York.

U. S. Treasury Dept., 1959, Tables of useful lives of depreciable property: U. S. Treasury Dept. Bull. F, 67 p.

Tillson, B. F., 1938, Mine plant: Am. Inst. Mining and Metall. Engineers, New York, 371 p. (out-of-print).

Uranium Prospectors' Handbook, 1954, Repro-Tech. Inc., 3535 Tejon St., Denver 11, Colo., 22 chapters.

Von Bernewitz, M. W. (Revised by H. C. Chellson) 1943, Handbook for prospectors of small mines: McGraw-Hill Book Co., Inc., New York, 547 p.

Willcox, Frank, 1949, Mine accounting and financial administration: Pitman Pub. Corp., New York, 489 p.

Young, G. J., Elements of mining: McGraw-Hill Book Co., Inc., New York.